GOLD RUSH
IN THE
JUNGLE

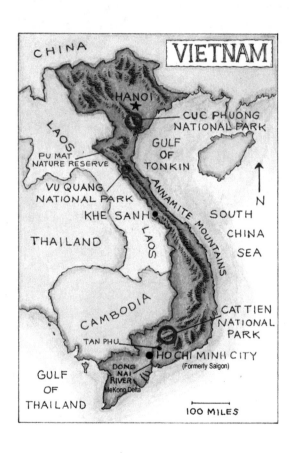

GOLD RUSH
IN THE
JUNGLE

The Race to Discover
and Defend
the Rarest Animals of
Vietnam's "Lost World"

DAN DROLLETTE JR.

CROWN PUBLISHERS · NEW YORK

Published in the United States by Crown Publishers, an imprint of the Crown
Publishing Group, a division of Random House, Inc., New York.

www.crownpublishing.com

CROWN and the Crown colophon are registered trademarks
of Random House, Inc.

Portions of the materials in the prologue and chapters 1, 5, 7, and 16
previously appeared in *Newsday, International Wildlife, The Sciences,
National Wildlife,* and *Scientific American.*

Library of Congress Cataloging-in-Publication Data is available upon request.

ISBN 978-0-307-40704-7

eISBN 978-0-307-95587-6

Printed in the United States of America

Book design by Maria Elias

Map by Patricia J. Wynne

Jacket design by Sara Wood

Jacket photograph: © Peter Unger

10 9 8 7 6 5 4 3 2 1

First Edition

To my parents, for instilling in me a love of nature
and a sense of curiosity about the world.

Contents

Author's Note

To see beneath both beauty and ugliness; to see the boredom,
and the horror, and the glory . . .

—T. S. Eliot

The discovery of a new species of wildlife is often the result of multiple strands of competing evidence, chance encounters, accidents, preconception-shattering experiences, and winding, discursive roundabouts. To even begin to describe the situation in Vietnam when it comes to discovering and rescuing new species, I attempted to conduct as many interviews as possible, with multiple sources, concentrating especially on the men and women who sweat it out among the mud and leeches.

To put these fieldworkers' immediate, firsthand observations into larger perspective, I sought the viewpoints of those who work at wildlife organizations, zoos, nature preserves, research centers, museums, and universities in other parts of the world, along with the printed record. I have tried to give the big picture vitality by relying upon an abundance of significant detail, telling this wildlife protection story in terms of people, places, and events as opposed to quoting scientific journals or government documents verbatim—although such publications have been priceless for background.

Whenever possible, I brought in the point of view of a newly emerging generation of young Vietnamese biologists, park rangers, graduate students, and PhD candidates with an interest in protecting their country's natural heritage. Their input is often overlooked due to the language barrier and the sometimes byzantine bureaucracy involved in working with the Socialist Republic of Viet Nam. (For clarity's sake and to be in line with conventional Western usage, I have otherwise written the country's name as "Vietnam.")

I chose to focus on wild mammals because that is where researchers have encountered the largest and most spectacular creatures. But I could have just as easily concentrated on plants, fish, insects, amphibians, or bird life.

Similarly, I zoomed in on a few key protection programs and parks scattered up and down the country out of the dozens of reserves available, paying most attention to Cuc Phuong National Park and its environs. As Vietnam's oldest and best-established nature preserve, since 1962 this place has some of the country's most innovative wildlife rescue programs, setting the tone for the rest of Indochina.

A few words to those wishing to see for themselves some of the rare and wonderful creatures described here. Parts of Vietnam are striking, but not much of it is in the kind of untouched state that Westerners are used to in places such as Yellowstone National Park. After all, Vietnam has been inhabited by successive waves of people over thousands of years.

Yet biogeographic islands remain, off the tourist trail. Even heavily visited places such as Ha Long Bay have rare and endangered langurs scurrying up and down the rock cliff faces of its islands. You may not be able to see a saola or a kouprey, but you could see one of the last sixty specimens of the highly endangered Cat Ba golden-headed langur.

One last bit of advice: Give yourself plenty of time, and put away the itinerary now and then. One highlight of my last trip occurred simply by sitting in a Hanoi café and talking to the proprietor and one of the regulars. I'd only planned to stop for a brief coffee before embarking on an ambitious two-hour walk to tick off every item in my guidebook's "Must-See" list of the city. Instead, a quick break turned into a fascinating three hours in which we exchanged pictures, looked at maps, and traded e-mail addresses; I was able to visit the newsroom of the *Vietnam News* as a result. I never did finish that walk.

Like the cramped, narrow streets of Hanoi's Old Quarter, the work of wildlife biologists in Vietnam is full of the unplanned and serendipitous, as well as the messy, the crowded, the dirty, and the wonderful.

This is their story.

As a reporter, I have tried to be as honest, accurate, intelligent, and fair as possible; any errors of fact or interpretation are mine.

Prologue

Dawn in the Jungle

This region represents much more than the find of the year;
it could be the find of the century.

—Colin Groves, taxonomist, Australian National University

It is daybreak in Ninh Binh province, seventy-four miles southwest of Hanoi, and the limestone mountains of Cuc Phuong—Vietnam's first national park, founded in 1962 with the blessing of Ho Chi Minh himself—are just emerging from the mist.

Though it is only five a.m., lights can already be seen in the windows of the farmhouses just outside the park; the buildings' traditional thatched roofs, combined with the adjacent neatly tilled rice paddies and abrupt nearby mountains, make the scene look as quiet and still as that on an ancient scroll.

Inside the park, however, the forest is full of sound, from the drone of mosquitoes to the maniacal racket of white-crested laughingthrushes. Loudest of all is a deep-throated "huuuu-huuuuuuu-huuuuuuuuuu-huuuuuuuuuuuuuuuuuuu" coming from the dense treetops. This is the "great call" of the gibbon, a long-armed, fruit-eating ape, which human listeners have sometimes compared to a mourning dove's cry, managing to be beautiful while mixed with a sense of loss.

Unfortunately, it is a sound fast disappearing from Vietnam's forests, at a pace that has accelerated noticeably over the past fifteen years.

One of the few places where you can still hear what Jane Goodall once described as "one of the wonders of the primate world" is here, just inside the park boundary, at the Endangered Primate Rescue Center (EPRC). Consisting of a five-acre semiwild, enclosed area, the center's roughly circular central compound lies inside a larger perimeter ringed by two outer fences; from above, the series of concentric circles would resemble a dartboard. And in the bull's-eye are 150 specimens of the world's rarest and most endangered animals, most of which would have been dead but for the efforts of Tilo Nadler (TEE-low NAD-ler)—a self-taught biologist who nevertheless went on to become what an eminent zoologist, Colin Groves, described as "the unsung hero of Indochina wildlife protection."

There is much to be protected, because just beyond the park lie a series of limestone mountain ridges stretching north to south to form the calciferous spine of Indochina. Known to geographers by the lovely name of the Annamese Cordillera, this mountain range runs the entire length of the country and contains within its valleys, hills, sinkholes, karsts, and innumerable caves something that Oxford University zoologist John MacKinnon described as "the lost world"—home to strange, rare animals such as the Asiatic sun bear, the Tonkin snub-nosed monkey, and the clouded leopard.

But that's not all. Since the early 1990s, many new, fantastic, large mammals never seen before by Western science have popped up here, especially in the parts where the borders of Vietnam meet those of Laos or Cambodia. Every week for the past ten years, an average of two new species of animal or plant have been found, all previously unknown to the outside world. A short list includes a

half-goat/half-ox, a deer that barks, a creature that may be a missing link between domestic cattle and their wild forebears, and a
close relative of the nearly extinct Javan rhino, to name but a few. A
previously unknown, new species of the very rare leaf-eating monkey known as the langur once appeared, quite literally, on Nadler's
doorstep. Others are still being discovered and are part of what the
peer-reviewed journal *Science* called "a renaissance in species discovery, not just of insects and microbes, but also of humans' closest
relatives, mammals."

There are also tantalizing, persistent reports by local villagers of
a fur-covered animal they call Nguoi Rung (NOW-rung), or "Forest Man," which walks upright on two legs and is said to resemble a
human. (The Nguoi Rung is most likely a myth, but no researcher
I talked to wanted to rule it out completely, bearing in mind what
happened with newly discovered species of previous eras. When
mid–nineteenth century explorers first heard accounts of a "hairy
man" in the mountains of interior Africa, experts discounted the
reports as mere fables. Now we know these creatures as Rwanda's
"mountain gorillas.")

How Vietnam's animals came to populate this "lost world,"
how they survived what locals call the American War, and how
they managed to still remain undiscovered—how can no one notice
a two-thousand-pound forest ox?—is a bit of a mystery. Whether
these species will survive the peace is still being decided, and the
outcome is very uncertain.

For as fast as these new creatures are being discovered and
formally described, they are being wiped out. Already, the newly
discovered Javan rhino is extinct in Vietnam, while the goatlike
saola has seen its numbers plunge from the thousands to approximately two hundred. The forest ox, or kouprey, may already be
gone as well.

The result is a race between the forces of preservation and destruction in this part of the tropics—the band of terrain where most

of the world's biodiversity is found. Researchers want to find and name the new species so they can take the creatures to rescue centers and captive breeding programs, and understand these animals' places in the great fabric of life. Meanwhile, others want to slaughter the animals to satisfy the newfound taste for exotic game in upscale restaurants that has gone hand in hand with a booming Asian economy. Endangered animals—both newly discovered and previously known—are sought on many fronts: their heads go to trophy hunters, their still-beating hearts used for the making of "snake wine," their horns for quack medicine, their brains for appetizers, their anal glands for the manufacture of some of the world's most famous and expensive perfumes.

At times, it seems that everything is being sacrificed on the altar of pell-mell economic development, an attitude that caused Alan Rabinowitz, formerly a zoologist at the Wildlife Conservation Society, to dub Vietnam "a miniature China on amphetamines."

The situation presents a conundrum to wildlife biologists. On the one hand, they are shocked by the rapid decline of species; the number of Vietnam's turtles lost each year to smugglers is measured by the ton. At the same time, wildlife biologists are thrilled just to find something new. In an era when it is big news to discover a new kind of "tube worm," the thought of finding and naming a new, large terrestrial mammal is just short of mind-blowing. By 1812, noted French naturalist Georges Cuvier was already lamenting that all the big, four-legged creatures had been found, leaving nothing new to discover. In a phrase now gleefully repudiated by wildlife biologists, Cuvier wrote: "There is little hope of discovering new species of large quadrupeds."

Consequently, if you're a young wildlife biologist and you want to make a name for yourself, you hightail it to this part of Southeast Asia.

However, this opportunity comes at great risk. Where scientists elsewhere worry about getting tenure, researchers here must dodge leftover land mines and winged antipersonnel "butterfly bombs" to do their field research. This is a place where the phrase "publish or perish" has a very literal meaning. Members of one expedition awoke to find tiger tracks circling their tents; their leader, Nate Thayer, said: "Our team's plane crashed on the return, our security mutinied and threatened to kill us all, half the team thought they were going to die after we encountered armed Khmer Rouge, others collapsed from sheer exhaustion from having no idea what it took to walk thirty miles a day in the jungle with no water, some demanded nonexistent helicopter medevacs . . ."

Drug runners, timber smugglers, and their like abound, their numbers increasing the farther you get from Hanoi and the deeper you go into the mountainous border regions. If you do cross the line and enter the neighboring hills of Laos and Cambodia, you have entered an area with a reputation like that of America's lawless old Wild West. In the late 1990s, the world's only captive specimen of a type of barking deer lived here in the private menagerie of a local jungle warlord, his prize guarded by his personal army of mercenaries armed with Kalashnikov AK-47 automatic rifles. Thayer—himself a seasoned expatriate American who knows these parts well—simply described the whole region as "a very bad area, full of very bad people."

But the pull of the new wildlife is too strong.

And academics are not the only ones rushing in. The jungle disease–resistant genes of a single forest ox could be worth billions of dollars to the world's domestic cattle industry, says Noel Vietmeyer, a National Academy of Sciences expert on biological engineering.

The result can be thought of as a "biological gold rush," simmering away largely unnoticed by the outside world, by expatriate Viet Khieu, and by many Vietnam urbanites.

Like all gold rushes, this one has attracted a motley cross-section of humanity: scientists, collectors, mercenaries, poachers, wildlife conservationists, local villagers, timber barons, smugglers, warlords, and reputed spies.

All nationalities seem to be represented. In addition to the Vietnamese, Laotians, and Cambodians, there are British birdwatchers, American war veterans, Danish journalists, French businessmen, Dutch park managers, East German former communists, and Indian/American scientists. There is also a smattering of those with no real fixed address—who in the bars of the newly refurbished colonial-era hotels of Ho Chi Minh City, née Saigon, have been overheard to refer to themselves, self-mockingly or not, as "expatriate scum."

These groups are mixed in with the ethnic hill tribes such as the Bru, the T'ai, the Gie-Trieng, the Yao, and about fifty others. Though outsiders tend to lump all these different tribes together as the "Hmong," the term more accurately refers to a single distinct ethnic group, say anthropologists. (The term "Montagnard" is a similar catchall, simply meaning "mountain people.") Just one group, the Yao, make distinctions between Red Yao, Black Yao, and Flower Yao, the colorful names referring to the type of traditional headdress worn by the women of each ethnic group. Each seems to speak its own dialect.

Amid such a tangle of languages, nationalities, cultures, occupations, and motivations, to merely ask a question and get a straight answer about the possible sighting of a new species is a challenge. One experienced zoologist with sixty years of fieldwork experience under his belt, George Schaller, lamented of this region: "When I'd ask local villagers 'Are there barking deer here?' they'd say yes—but it turns out there's four kinds of them. I wasn't asking the right questions."

Under these circumstances, for Nadler's Endangered Primate Rescue Center to locate healthy, adult pairs of rare, newly discovered animals and then get them to mate and raise live young in captivity is nothing less than astonishing. Successfully breeding animals in confinement is considered to be the ultimate test of the skills of any deep-pocketed, resource-rich big-city zoo, let alone those of a small volunteer operation working in the hinterlands of a Third World country. What Tilo Nadler did with just one creature, a leaf-eater known as the Ha Tinh langur, was "unprecedented," said Colin Groves, a renowned mammalogist and taxonomist at the Australian National University who is an expert on endangered primates. "The Ha Tinh langur has never been in captivity before. Tilo has chalked up quite an achievement."

And Nadler's EPRC has had similar results with the even rarer Delacour's langurs, Cat Ba langurs, and many other species of leaf-eaters, while spawning similar rescue efforts among other species in Cuc Phuong National Park and the region as a whole.

For there is much wildlife rescue work going on in Vietnam these days. A short list of some of the many wildlife discovery and restoration projects in recent memory would have to include Gert and Ina Polet's effort to save the country's previously unknown Javan rhinos near Saigon, Jeb Barzen and Tran Triet's work to bring back Eastern Sarus cranes on the Plain of Reeds, William Robichaud's expeditions to find the saola in the hill country, and Bui Dang Phong and Tim McCormack's work in saving the world's rarest turtles, at least one of which lives—of all unlikely places—in the middle of downtown Hanoi.

To this list must be added earlier expeditions such as Charles Wharton's agonizingly close near successes in catching a live kouprey in the early 1950s, or Vo Quy's effort in the 1960s to make a complete inventory of all the bird life in the country—the first-ever such census, and conducted along a major weapons-supply route, in the middle of a war.

The list of creatures involved goes on and on, to include paddlefish, raccoon dogs, fishing cats, warty pigs, and "ferret-badgers," plus innumerable other creatures strange to Western eyes. (Some of these animals, such as the ferret-badger, are so unusual that taxonomists are still sorting out where they fit in and what their proper names should be. They are so new to science that they cannot be found in field guides or textbooks or encyclopedias yet.)

But in the dawn of the twenty-first century, the one project that seems to stand out whenever an outsider penetrates the small community of wildlife rescuers is Tilo Nadler's Endangered Primate Rescue Center. The EPRC is one of the oldest efforts in this country to actually rescue rare wild animals as opposed to strictly observing them, and it started in the early 1990s, placing it among one of the earliest peacetime restoration projects.

My curiosity had been piqued early and often by this effort. I had heard rave reviews of Nadler's work from widely scattered researchers in many different disciplines in several different countries; the most colorful accolades came from mammalogist and taxonomist Colin Groves, when I visited his offices at the Australian National University (ANU) in Canberra.

It was a memorable interview. Exotic souvenirs covered his office walls; we were surrounded by animal heads, maps, and photos of him posed with creatures in faraway lands. It felt like being on a movie set; one half expected to see a leather jacket and a bullwhip.

During the obligatory Australian teatime, Groves filled me in on the hottest developments in his field. He described a frenzied but little-publicized series of new biological discoveries going on in Vietnam's jungles, where researchers had recently found some brand-new species of large, four-footed animals that did not easily fit any previously defined categories.

While the idea of all these new species was stunning, he said, even newer ones seemingly popped up every time someone installed a camera trap. What's more, there were tantalizing hints—tracks, dung, bits of antler, leftovers in village cooking pots—of many more creatures yet to be formally discovered, named, and categorized. Capturing any one of them would make the career of an eager biology grad student.

Or a trophy hunter.

Groves said that Vietnam's wildlife had managed to thrive and remain unseen by a series of remarkable twists of fate. By way of an example, he pulled out an old atlas of Indochina sandwiched between books on his office shelves that carried titles like *Congo Journey* and *New Conquest of Central Asia*. Drawn upon the yellowing map were the routes of the only major scientific expeditions to survey Vietnam's wilderness in the twentieth century. Groves said that there were only two significant countrywide, all-encompassing forays, both in the late 1920s, toward the peak of the French colonial empire in Asia. That was the era of the great white hunter, when expeditions traveled by elephant back or sedan chair while accompanied by immense trains of hundreds of porters—expeditions whose very size restricted their mobility in the confined spaces of the mountainous jungle. Convoys of 250 porters were needed just to carry one week's supplies to the hunting lodges in the popular big-game safari region of Dalat, in the subalpine hill country northeast of Saigon. A single party of a handful of Western tourists required fifty native porters—who were often unpaid. It's little surprise that a surviving handbook from the era included such phrases as "How many coolies can this village provide?" and "Make a straight path, not a crooked one." Or "I don't want children, I need vigorous men."

Tracing their imperial paths with his fingertip, Groves showed how both expeditions just missed the richest and most ecologically diverse regions, due to the poor roads, severe conditions, prolonged rainy seasons, and endemic tropical diseases. In the earliest days of colonization—the 1860s—European administrators, explorers, and military men suffered horrifying mortality rates; one doctor estimated that the average European only survived two years in

Vietnam. In an era before science understood the connection be-tween malaria and mosquitoes, medical practitioners were helpless: a Dr. Fontaine, who was the equivalent of the surgeon general for the entire colony of Indochina, himself died from tropical disease between the time he submitted a paper on the topic and the time it was published.

With new drugs, the mortality rate was reduced dramatically, but certain regions continued to remain unknown or largely for-gotten. Groves lingered over one spot, adding: "This is where Tilo Nadler and the Endangered Primate Rescue Center are. He's doing the most exciting work today to save rare animals."

That was it; I had to go there. I traveled to Vietnam to meet Nadler and others for the first time in 1998. I was consumed by questions: Who were these renegade biologists? What was Nadler's Endangered Primate Rescue Center doing that made it so success-ful? And if Nadler's work was so wonderful, why had no one outside a small coterie of wildlife experts heard of it? What was happening in Vietnam? What would I find when I met them at their jungle field stations?

My first visit left me with more questions than it answered. I decided to return more than a decade later to try to answer them once and for all.

1

A Peace More
Dangerous Than War

It's a Cruel, Crazy, Beautiful World.

—Title song of album by South African
musicians Johnny Clegg & Savuka

IN 2011, CAREFUL newspaper readers might have noticed a small
article tucked away in the science and environment sections, which
said that authorities had determined that sometime during the previous year, Vietnam's last Javan rhino had been killed by poachers in Cat Tien National Park. Rangers had found the enormous
bullet-riddled corpse with its horn sawn off. It had taken more than
a year to determine absolutely that this had been the very last rhino
in Vietnam, and that there were no others hiding in the bush. A
typical story, such as the one on the BBC News website, noted that
this individual had been the last surviving member of the subspecies
Rhinoceros sondaicus annamiticus.

Dry and to the point, the news was a stunning surprise to anyone who knew the background. It was a shocking example of what

had been occurring in Indochina since I was last at Cat Tien in person, in 1998, to meet the wildlife biologists working there to rescue the animal from extinction.

It had been late in autumn monsoon season in the south of Vietnam that year, and the wet, humid weather had brought out armies of leeches. The wormlike creatures seemed to wait for us on the tips of nearly every jungle leaf in this young national park, where they would attach themselves to any warm-blooded creature that passed by.

Our little group constantly brushed leeches off our shirtsleeves and pants cuffs, only to discover telltale patches of blood on our socks after we stopped at a ranger's hut to dry off. When we sat down to gratefully drink some of the green tea the ranger offered, a leech began crawling across the tabletop toward us. We watched the leech inchworm its body among the teacups until park overseer Gert Polet could stand it no more and flicked the leech away—only to discover two more leeches had arrived.

Outside, the rain was getting worse. In some places, two feet of water covered the main hiking trail. Even the local guides refused to go outdoors. We were ninety-four miles northeast of Ho Chi Minh City (formerly Saigon), in an isolated area then known to few outsiders. The nearest landmark that appeared in any printed tour guide was the abandoned bunker complex at Cu Chi, now a tourist destination, located about halfway between where we were and the former capital of what had been South Vietnam.

The only way out of the park had turned into a trough of red mud. When I had ridden in earlier on the backseat of a motorcycle, the bike had sunk up to its wheel hubs in the stuff, which had been churned to the consistency of pudding by the street traffic at busy intersections. Even getting back to the nearest village was now questionable, since the Dong Nai River on the park's border was so high and turbulent that the ferry—actually a landing craft left over from the last war—had stopped operating.

It was a miserable situation, made worse as I was at the time still recovering from a weeklong encounter with a tropical illness picked up near Hue in the center of the country. (Coming down with amebic dysentery in the hill country seemed to be almost a rite of passage. I could only be thankful that it was not malaria; one researcher told me he had caught malaria sixteen times while in Indochina. I could now really appreciate what Groves had said about disease acting to keep outsiders away.) I was still shaky, and the constant, unrelenting monsoon rain only worsened the mood.

Despite the conditions, Phil Benstead and his girlfriend, Charlotte, from the Royal Society for the Protection of Birds, had broad smiles on their faces. Here to survey Vietnam's birdlife, they'd seen an enormous variety of winged creatures living in the second-growth forest of this former Viet Cong staging area.

"We identified a hundred and twenty different species yesterday, including edible-nest swiftlets. We even saw a gray-faced tit-babbler," said Phil in his proper English accent as he lit up a cigar to celebrate. He and Charlotte then discussed plans to look for owls in the jungle by flashlight that evening. I had to admire their indomitable cheer and pluck.

After they departed, I asked Polet whether he was concerned about the pair going out at night. Polet, then an adviser on park management at Cat Tien for the World Wildlife Fund, replied that he only became really anxious when people started poking around the dense thickets of young bamboo and rattan that had sprung up in place of the original old-growth forest. The thick tangle made it hard to see what was underfoot.

"Parts of the park still have unexploded bombs and land mines," the lanky blond, blue-eyed Polet explained, with a trace of his native Holland in his speech. "Whenever I send teams out there to do biological surveys, I just hope everyone gets back."

Leeches, land mines, and disease were just part of the reason that so much of Indochina's wildlife had remained undetected for so long. Warfare, mosquitoes, impassable terrain, and improbable quirks of fate had conspired to keep Vietnam's jungle creatures unknown to the modern world. Isolation—together with resiliency and luck—had enabled the country's wildlife to remain alive and undiscovered. But after suffering through decades of war, embargoes, and currency controls, Vietnam was now starting to open up more to outsiders, making one of the world's ecological hot spots accessible to researchers.

Much of the interest in the particular hot spot known as Cat Tien was due to a large, irritable, deceptively clumsy-looking animal. In 1988, a solitary rhinoceros had been killed near here; further research established that it was genetically related to the Javan rhinoceros, one of the most endangered mammals in the world. Before that discovery, the Javan rhino—which once ranged as far as India—was thought to be extinct on the Asian mainland and to have dwindled to fewer than sixty animals, all confined to a single preserve in the western part of the island of Java. A 1998 study of footprints suggested that possibly a dozen or more rhinoceroses still lurked in the northern reaches of Cat Tien, and zoologists seemed giddy at the thought. Investigators examined rhinoceros dung, made plaster casts of rhino tracks, and snapped photographs with automated cameras in an effort to determine what steps should be taken to preserve the species.

Some thought there was a chance to rebuild the rhino population in Vietnam, a goal that had some precedent. The population of Indian rhinos had been down to as few as twenty individuals before recovering to about five hundred. Similarly, at the end of the nineteenth century, there were only twenty or forty southern white rhinos in Africa; their population has since rebounded to more than eight thousand individuals.

But Polet was pessimistic. He noted that what Vietnamese rhinos his rangers had found were showing signs of stress and poor diet. The animals were physically smaller than their kin elsewhere, possibly due to the bamboo and rattan having crowded out the animals' usual fodder of small trees, shrubs, and grass. The size of the rhino sanctuary was still small for such a wide-roving animal, the nearby human population was expanding rapidly, much of the forest was illegally being cut down for fuel, and he suspected that the Javan rhinos' numbers were actually closer to five or seven individuals than to twelve. As it later turned out, he was right about the rhinos' chances, although no one then knew how desperate the situation was.

Upon hearing the news of the loss of the last wild Javan rhino in Vietnam in 2011, the chairman of the Asian Rhino Specialist Group of the International Union for the Conservation of Nature (IUCN), Bibhab Kumar Talukdar, was to tell the BBC that the animal's passing was "definitely a blow" to the survival of all Javan rhinos everywhere.

Others were blunter. Alan Rabinowitz, a zoologist at the Bronx Zoo once described by *Time* magazine as "the Indiana Jones of wildlife protection," called the demise of the last Vietnamese specimens "the dumbest thing." He said that a captive breeding program would have been the best way to go.

Rabinowitz said, "The last population of Javan Rhinos in Indochina is no way as strong and healthy now that that subspecies population in Vietnam has gone. They should not have let that last one or two Javan Rhinos stay out there. It was stupidity, pure stupidity."

Equally upset at the news was noted zoologist George Schaller of the Wildlife Conservation Society. Schaller has spent six decades trotting the globe to protect wildlife—including species as diverse as mountain gorillas, giant pandas, tigers, Serengeti lions, snow leopards, and Tibetan tigers—and he had personally been in Vietnam to help survey the populations of Javan rhinos and kouprey in the early '90s, spending months on end in the Indochina jungle. (When Schaller did fieldwork in Tibet, he was accompanied by a youngish Peter Matthiessen, who went on to write a literary classic about the experience, *The Snow Leopard*. George Schaller is the "GS" referred to throughout that book.) Now, a little more than two decades later, the Javan rhino was gone forever from Vietnam, a victim of greed and a rapacious black market. "There's tens of thousands of dollars in a single horn," Schaller lamented. "Consequently, rhino poaching is like the drug trade." One gram of rhino horn sells for as much as $133, or double the price of gold. Given the size of most rhino horns, that translates to an average of about $250,000 per horn.

The extinction of this animal was all the more disturbing because this one subspecies of rhino had been the justification behind creating the entire 270-square-mile national park in the first place. (Now that it has been up and running for close to twenty years, this well-established reserve is likely to remain. But nothing can be taken for granted; even the oldest park in Vietnam was cut in two to accommodate a new superhighway a few years ago.) As a champion example of what zoologists call "charismatic megafauna," the Javan rhino had attracted huge amounts of attention, support, and money. Consequently, the species' discovery had meant that one of the largest tropical rain forest lowlands in Vietnam had been getting more comprehensive, all-embracing protection. In order to protect rhino habitat, three smaller, widely spaced, fractured pieces of lowland forest had been joined together to form Cat Tien National Park.

Though composed of land that had been repeatedly sprayed with Agent Orange by American forces during the war, with large swaths of any remaining old-growth forest heavily logged afterward, the new preserve had proven to be home to hundreds of plant species, 120 kinds of birds, and several other large mammals, including the elephant, the sun bear, the buffalo-like gaur, and about forty others on the IUCN's "Red List" of endangered species.

But of them all, the Javan rhino had been the star, as closely identified with Cat Tien National Park as the grizzly bear is with Yellowstone. "Big, beautiful animals draw public attention," observed Schaller. "They arouse emotion, and people will work hard to protect them. Whereas it's difficult to arouse people to protect leeches, mosquitoes, and so forth—although they may be just as important to the overall ecosystem, or even more so. . . . Is a tick more ecologically important than a tiger? Ecologically speaking, we just don't know."

The reality, Schaller explained, is that while he and other zoologists can do scientific research about all the different species and their web of interrelationships to urge the establishment of a national park, conservation often boils down to politics and emotions. People can relate more easily to any kind of large animal; there was a different energy to a place where big wild creatures roamed, and a sense of wilderness that came with the presence of a rhino that cannot be re-created digitally. There was something to being in the forest and knowing that you were not the biggest animal there. It sent a shiver down the neck.

In addition, there was a bonus to protecting the most glamorous animals. "If you have an appealing animal like a rhino or panda or tiger—and they need a fair amount of space—then when you do protect a good population of them in their natural habitats, you automatically protect hundreds or thousands of other animals and plants as well," Schaller stated.

Rabinowitz described such animals as "apex species," saying that by saving the most appealing animals and the land they migrate over, ecologists could also save the less noticeable creatures associated with them. This meant that by publicly focusing an enormous, concentrated effort on one species, they could save a whole stable, rich environment with a cascade of other life-forms in the bargain.

In this tapestry of protected animal and plant life, Vietnam's Javan rhinos were the "apex species." As the primary focus of all the attention, they simply should not have disappeared.

1998

2

An American in Vietnam

"Tay ba lo" means "Westerner with a backpack."

—Claude Potvin and Nicholas Stedman,
Dos and Don'ts in Vietnam

FINDING OUT WHY some wildlife restoration projects fail and some succeed can be harder than anyone would first think. Merely to extract some information can be frustrating now, but it was much worse in the pre-Wikipedia, pre-Google era. It was extremely difficult to find any data of any kind about either the creatures or contemporary Vietnam, whether online, in print, or anywhere else in 1998.

Vietnam then was just emerging from decades of isolation, and the United States had only formally recognized Vietnam the year before, in 1997. And that was a different time in the journalism game, before the rise of blogs, RSS feeds, SMS messaging, social media, podcasts, live chats, and Twitter—even if the content today is often censored by the Communist Party authorities in Vietnam.

Making it even harder, field biologists have always had a tendency to get so wrapped up in their work that they often simply do not respond to queries by e-mail, fax, or phone call. As a group, they give the impression that they'd much rather be out in the bush among animals than answering questions from nosy outsiders. As one researcher apologetically offered, "We don't often like to mix that much with people."

By their own account, wildlife biologists can be a driven, eccentric lot. Oxford University biologist John MacKinnon, among the first to formally discover Vietnam's saola, blamed this on the loneliness of fieldwork, difficult conditions, and endless months with no one to talk to. MacKinnon told a reporter: "I knew of a couple, husband and wife, who did fieldwork together. They went quite mad in the forest, talking to inanimate objects. They gave everything names. The kettle was called Billy, as in 'Billy the kettle, are you ready?'"

To put up with these hardships requires an obsession similar to that of mountain climbers or early polar explorers. (In her nonfiction book *The Orchid Thief,* Susan Orlean described how nineteenth-century orchid collectors would even dig up graveyards to collect a prized flower. She wrote, "Because orchid hunters hated the thought of another hunter's finding any plants they might have missed, they would 'collect out' an area, and then they would burn the place down.")

Part of this fanaticism may be laid to the human tendency to want to best one's rivals, and, as such, it has been going on since time immemorial. The competition between rival paleontologists Othniel Marsh and Edward Cope in the American West in the 1870s is legendary among historians of science today. These two respected scientists, backed by well-known institutions, frantically raced each other to be the first to collect new fossils of previously unknown dinosaurs—including the first *Apatosaurus, Camarasauras, Diplodocus, Stegosaurus,* and *Allosaurus*—in places such as Wyoming's Badlands. They started their excavations there mere weeks after Custer and his entire Seventh Cavalry had been wiped out at the Battle of the Little Bighorn in neighboring Montana.

The ensuing, decades-long series of battles between Marsh and Cope, sometimes called the Great Dinosaur Rush, was scarcely less intense than Custer's Last Stand. Mutual accusations flew back and forth about sabotage, bribery, fraud, theft, and spying. The rival paleontologists' teams worked in the depths of winter, when temperatures plummeted to minus thirty degrees Fahrenheit, or minus thirty-five degrees Celsius, *before* any windchill factor, in an effort to be the first to excavate a new species and claim the naming rights. Tons of fossilized bones were sent to museums back east, in a scientific competition run amok.

Each was in such a rush to beat the other into print that modern scientists termed their efforts "taxonomic carpet bombing." In some cases, dozens of different scientific names were applied to what turned out to be just one species. It took decades to sort out the mess.

Worse, if Marsh came across a dinosaur fossil that he could not remove, he told his team to destroy the bones rather than see them fall into the hands of his hated opponent. One of today's best-known paleontologists, Bob Bakker, curator of paleontology at the Houston Museum of Natural Science, described this attitude as the "Fred C. Dobbs Syndrome," after Humphrey Bogart's crazed gold prospector in *Treasure of the Sierra Madre*.

Legitimate scientists today still push themselves to incredible lengths to conduct their work. While he was still a young man and had yet to write any of the books that would later make him famous, the evolutionary biologist Ernst Mayr traveled to Papua New Guinea in the 1920s to collect bird specimens for the American Museum of Natural History. In those days, collectors gathered birds in the same way that John James Audubon had a century earlier: they shot them. Audubon considered a day in which he shot fewer than a hundred birds to be a wasted day; he once described himself as "a two-legged monster armed with a gun . . . leaving a path of destruction."

After the killing, naturalists then gathered up the skins, feathers, and bones to make stuffed mounts for museum collections and displays. But rather than throw out the internal organs and meat from his bird specimens, Mayr told me that he ate them in the field; the bird carcasses were one of the few sources of protein he could find in the Papua New Guinea jungle in the era prior to freeze-dried food and Meals Ready to Eat. It was a case of eating your research subjects to survive.

Mayr remembered the circumstances vividly—even though he was nearly ninety-seven years old at the time of our interview—and laughed at the recollection of his earliest fieldwork, saying that it added a twist to the old academic saw "publish or perish." (Mayr ultimately did much more than just publish, becoming the dean of the biology department at Harvard University, where he introduced key concepts to our understanding of how new species come about, such as "island biogeography." When he retired, the university's zoology museum was renamed in his honor. After our interview, Mayr was to write one more book on evolutionary biology, his twelfth, before dying at age one hundred.)

All of which indicated that field biologists were sometimes consumed by their work in the bush. It was a variation of something I had encountered several times while working as a foreign correspondent in Australia for a science magazine. Once, while trying to do a story on a prestigious marine mammal biologist in the remote tropics of the land down under, I found that this person had gotten so caught up in the tropical lifestyle that he had essentially dropped off the map, living like a modern-day castaway. This kind of thing happened often enough that the Aussies had an expression for it: "gone troppo." (Translation: "gone tropical.")

It was frustrating, particularly if a deadline was coming up fast; the more you pushed, the slower the response. At times, it felt as if they deliberately slowed down out of spite.

Paradoxically, this had the effect of making the researchers in Vietnam more appealing and intriguing. Even today, some of the fieldwork happening there is a bit mysterious, even to the most authoritative sources. In 2012, renowned zoologist Alan Rabinowitz would say, "I've read about Tilo's primates, which I've heard about for years, but didn't think about it much. Often, a foreigner comes in, saves a few creatures, and then not much more happens. But from the bits and pieces I could gather, Tilo seems to be doing things right: mentoring, thinking long-term, going into the community, building the next generation of wildlife rescuers to take over. It doesn't seem to be a Dian Fossey–type situation, where a lot was for an agenda, and there was a lot of rancor with the local people. . . . I have only been able to find out a bit about him, but I'd really like to meet him and find out more."

He added, "I'm very impressed."

If the researchers in Vietnam were mysterious and exotic, then the land in which they worked was more so. As Maxwell Taylor, US ambassador to Saigon during the height of the American involvement, had said in the 1960s, "We never understood them [the people of South Vietnam], and that was a surprise. And we knew even less about North Vietnam. Who was Ho Chi Minh? No one knew."

The situation only grew worse after the war, when the country was more or less closed off to the outside world, a complete unknown. Even in the '90s, it was still a cipher.

In addition, Vietnam was—and is—more than a bit rough-and-ready: Westerners riding on the *Reunification Express* were often surprised to discover that the toilet was just a hole in the metal floor of the moving express train. There was no seat, and if you looked down between your legs, you could see railroad ties passing underneath. A plastic bucket filled with water took the place of toilet paper.

The phrase "culture shock" didn't begin to do the place justice.

A full-time professional travel writer in New York City named Beth Livermore characterized Vietnam as a place with a reputation as a "hard country" for travelers.

Of course, that was part of the appeal—it had not been "discovered." Even at the time of this writing in the twenty-first century, there is not a single McDonald's in the country.

Nor were there many reporters covering Indochina in the '90s. With the land at peace and an array of difficulties in obtaining official permissions from the Communist Party authorities, there was little incentive for a reporter to go to Indochina.

One of the few exceptions to this rule was a journalist named Nate Thayer of the *Far East Economic Review* in Cambodia, who had years of experience in the region, an enormous number of contacts, and some amazing anecdotes. With a shaven head, T-shirt, jeans, and a wad of chewing tobacco constantly in his mouth, this son of an American diplomat cut a larger-than-life figure.

He also had a stellar reputation. Alan Dawson, editor of the *Bangkok Post,* wrote: "In my opinion, which is shared by many colleagues, Nate is simply the best reporter to come to the Indochina scene since the fall of Saigon in 1975."

In conversation, Thayer seemed to relish the murky demi-world in which mercenaries, secret armies, and the like moved about in Indochina, where it was sometimes hard to tell who were the good guys and who were the bad guys. His experiences exemplified the risks taken by those who go looking for wildlife in the region, be they recognized scientists or quasi-legitimate "collectors" of specimens. You never knew what you were getting into.

Thayer's most recognized Indochina coup read like an adventure novel: "After a series of furtive rendezvous, using coded messages over mobile phones, I slipped into one of the most impenetrable, malaria-ridden and land mine–strewn jungles of the world: Khmer Rouge–controlled Cambodia. . . . All of a sudden, there we were with Pol Pot, with no warning, no notice, in what was probably the only trial, or pseudo-trial, we will ever see of one of the world's most notorious mass murderers."

After that experience, it seemed natural to Thayer to mount a twenty-six-person expedition by elephant back in 1994 to capture a living specimen of the kouprey—the holy grail of zoologists ever since a live male calf was found by chance in a Paris zoo in 1937, when zookeepers discovered an unusual new species among a shipment of wild animal specimens from French Indochina. Unfortunately, it died a few years later, and not one has been in a Western zoo again. The animal, *Bos sauveli,* was later described by the author of a National Academy of Sciences publication as "perhaps the most primitive of living cattle," with features little changed since the Pleistocene era some 600,000 years ago. That same author, Noel Vietmeyer, calculated that the kouprey's inherent resistance to deadly diseases such as rinderpest would make it worth billions of dollars to the domestic cattle-breeding industry, if its genes could ever be introduced into the bloodline. Those figures were a good part of the reason for the interest in the animal.

But all efforts since that time to find and catch a live kouprey in the wild and transport it to a Western zoo have been dogged with disaster. The failed efforts are near legendary in Indochina, composed of a litany of guerrilla ambushes, land mines, malaria, mutinies, and other mishaps.

Despite all this, one zoologist, Charles Wharton of Georgia State University, had actually filmed a small herd of kouprey in the wilds of Cambodia in 1952, and even held three of the animals overnight in pens on separate occasions. Each time, they escaped. Then, the escalating war between France and Vietnam put a stop to further searches; it would be decades before anyone would try again, despite unconfirmed sightings reported in local newspapers.

"It's almost like the thing has some sort of ancient spell over it that man is not to learn about or capture this animal," lamented Wharton.

Predictably, Thayer's 1994 expedition was not a success.

Thayer himself called the trip, which he organized, "ridiculous." He added, "There were several early casualties from heat prostration and other manifestations of badly out-of-shape bodies addled by long histories of drug and alcohol abuse." As he recounted later, "These are very bad areas full of very bad people where no authorized environmental groups are likely, or willing, to go." When asked about the experience, he told me that he would never do it again—finding Pol Pot was apparently less stressful than finding a kouprey.

However, almost as an aside, Thayer added that he had actually eaten kouprey meat with some Cambodian villagers—albeit not on that particular expedition—and that he was personally convinced that the animal was still extant as of 2012 in the northeastern province of Mondulkiri, Cambodia, not far from the border with Vietnam.

A pithy note summed up his take on the animal's status:

It was Khmer Rouge territory when I was there.
We ran into Khmer Rouge patrols who were not happy
　　we were there.
Our security guards refused to accompany us.
Our trackers were scared shitless.
VN can't go because it is in Cambodia.
Cambodians don't go because there is no money to be made.
Traditional wildlife people don't go because it is too dangerous.
That is why the kouprey goes there.
Because nobody bothers them.

Thayer also called the funding for wildlife protection in Indochina "a racket."

But our story is getting ahead of itself.

. . .

While certainly not encountering anything like the Khmer Rouge, I did run across disturbing aspects of Indochina on that first extended trip into the field, in which moments of great beauty were matched by equally chilling ones. The long, nearly empty tropical beaches of Vietnam were a delight, as were the food and the friendliness of the local people: in a remote coastal village one day, an old villager picked up her grandchild and waved the child's arm at the foreigner while saying hello, or *"Xin chao"* (pronounced "Sin chow"). This was to happen many times; I was also invited out to meals and into people's homes.

In contrast, an urbane, multilingual Vietnamese man aboard the *Reunification Express* once gave a precise, step-by-step set of instructions about how to prepare the brains of live monkeys for dinner—a specialty of Lang Son province, near the Chinese border. My informant, Mr. Huyen, said that the meal was prepared by placing the tightly bound animal in an upright position and inserting its head into a circular hole fashioned for that purpose in the center of a dining table. (The table itself was just high enough off the floor to straddle the animal, so that the crown of the monkey's head stood just a little above the level of the tabletop.)

The skull of the still-living animal was then sawn open like a tin can, and thin slivers of its brain shaved off a layer at a time, dipped into a glass of wine, and eaten raw. Huyen said, "I do not like this thing," and made a face as he described it, which was officially outlawed but still occurred. (Corroborating the practice, the Vietnamese girlfriend of a wildlife biologist said, "The brains are the best-tasting part of a monkey.") When pressed, Huyen denied participating himself, but said that he had "heard about it from a friend."

Another disturbing feature to Vietnam was the war, which was still in everyone's minds, at least for Americans—even to someone who had only watched the fighting on the television nightly news as a child at the time. Partly because of the sheer size of America's war in Vietnam, and partly because it was still recent enough, the Vietnam War—or what historians call the Second Indochina War—formed a small, steady, constantly running drumbeat in the background, making Vietnam into a place of strange, surreal vignettes decades later.

Peddlers on city streets hawked Zippo lighters, helmets, and "genuine" dog tags of lost American soldiers at two dollars apiece. The wreckage of American warplanes lined museum halls. One of the most popular tourist attractions was the former Viet Cong guerrilla hideout of Cu Chi, in the old "Iron Triangle" a few miles outside of what is now Ho Chi Minh City, where modern tourists paid admission to crawl through former enemy tunnels. For an extra fee, they could take some target practice with an old M16 at a nearby rifle range.

The country's most popular chain of bars, decked out in a 1960s war motif, operated under the name Apocalypse Now. As Vietnam eagerly opened its doors to foreign investment and Western tourists, the entire country seemed to be a funhouse version of the conflict known here as the American War.

There were many, less immediately obvious, reminders as well: when flying into Hanoi, among the first things passengers saw were roughly circular pits that broke up the symmetrical, rectangular pattern of the rice fields. These were old bomb craters that had since filled with water; villagers used them to raise fish and ducks.

Visiting an archaeology museum in Da Nang devoted to the thousand-year-old red sandstone sculptures of the vanished, India-derived Cham culture, I learned that soldiers once slung their hammocks between the statues in the 1960s, before going into battle.

A hint of an earlier war—the First Indochina War, fought between France and Vietnam—came when visiting the fully restored, century-old wedding cake of an opera house in Hanoi: there were still bullet holes in one wall, left over from the start of the rebellion against the French Empire in September 1945. A plaque treated these scars much like we Americans treat the crack in the Liberty Bell.

But "our war" still colored America's perceptions of the present-day country of Vietnam, even after the passing of so much time. In fact, even in the second decade of the twenty-first century, Martha Hurley, one of the authors of *Vietnam: A Natural History*—one of the first sweeping English-language compendiums of Vietnam's wildlife—found that whenever she opened the floor to questions after giving a biology lecture on Indochina, the first thing asked was always about the Vietnam War.

In contrast, among the Vietnamese, there was not much agonizing over the war between their country and America. Instead, they went to great effort to emphasize the positive side of the history between the two countries, such as the fact that when Ho Chi Minh announced the start of his war of liberation against the French near the front of the Opera House in 1945, he quoted the opening lines of the American Declaration of Independence, word for word. Or that Ho's fighters had worked side by side with US forces to rescue downed American pilots in World War II, and that the precursor of the CIA had helped train his men to fight the Japanese Imperial Army.

And as wildlife biologists such as Polet were to point out, nowadays, among the Vietnamese, the problems seemed to stem more from who had served on the North Vietnamese/Communist side and who had been on the South Vietnamese/US side. "Sometimes when the guys get to drinking at night, the old stories come out, and you can feel the tensions rise," Polet said.

Another biologist, Jeb Barzen, working to restore endangered birdlife on the Plain of Reeds, southwest of Ho Chi Minh City, made the same observations. He said simply: "Civil wars are always like that. Our own Civil War is still a sore point among Americans to this day, and that was over a hundred years ago. Imagine what it's like for the Vietnamese."

And according to Robert Templer, the author of *Shadows and Wind*—an account of modern-day Vietnam—among the Vietnamese people, a popular stereotype was that South Vietnam had been populated by Vietnamese collaborators who cashed in on the boom times while the US troops were bivouacked there, rather than fight. Meanwhile, the image of the postwar North Vietnamese was that they had treated the South like a horde of carpetbaggers intent on plundering the region. Certainly, reporter Stanley Karnow wrote about the wives of Northern generals filling up military planes with loot purchased dirt-cheap from desperate, impoverished Saigon-dwellers after the North's victory in 1975. Both perceptions still raise strong resentments to this day.

But as Western scientists in a very foreign land, the biologists tried to steer clear of it all.

What Barzen—an American biologist in Vietnam on and off since 1988, under the auspices of the International Crane Foundation—learned was that because of those tensions, there were certain things that he could not accomplish as a foreigner, and neither could any Northerners. It turned out that the only person who could set up a nature preserve in the Plain of Reeds was a Southerner, born and raised in the area, who had spent a lifetime hiding in the dense foliage of its marshlands and fighting outsiders of any stripe as far back as World War II, when he staged ambushes against the invading Japanese. A tough old guerrilla, this Southerner—named Nguyen Xuan Truong—became the driving force behind preserving the wetlands he had come to know and love, back at a time in the 1980s when every scrap of land was being taken over for rice production, willy-nilly.

Nguyen Xuan Truong went on to become the leader of the local government and the founding father of Tram Chim National Park, as well as "the best ecologist in the Plain of Reeds I've ever seen," commented Barzen. "Nguyen could dip his finger in the water, taste it and give me the pH [the level of acidity]. I tested him, and he was right. He had lived there for decades, and he was a good observer—very knowledgeable, who understood exactly the hydrology of the area."

Barzen added, "Even though Nguyen isn't a formally trained scientist, I feel that our job is to 'translate' his knowledge, test it, and use it as the basis for restoration."

These wartime-related twists to the wildlife's story seemed to crop up regularly, and as a visitor they were the first things I became aware of: the ex-guerrilla who was an amateur hydrology expert, the warlord who was a nature lover, the guide who grew up in the old demilitarized zone, the biologist who had done bird surveys on the Ho Chi Minh Trail during its heyday as a weapons supply route, the ex-infantryman's story about spotting a Forest Man while on patrol. It may be that first-timers to Vietnam inevitably focused on the Vietnam War initially, and only later became more aware of the country's other features. Or it may be that in the 1990s, America's war in Vietnam was that much closer, and time had yet to heal many wounds.

After so many years have passed and so much has changed, it is hard for the average American high school student to understand what a grip the Domino Theory once held on the imagination of American policymakers during the Cold War—the idea that if Vietnam fell, the rest of the countries of Asia would fall to the Communists as well. The United States had picked up where the French left off in Vietnam, albeit with different motives: while the French wanted to hold on to their empire, we got into a war in Vietnam with the idea of fending off the intentions of the Soviet Union and Mao's "Red China"; the latter proclaimed itself to be as closely allied to Vietnam "as lips to teeth."

Today, the Soviet Union is no more, Mao is long gone, and China is hardly Red in the way cold warriors would recognize. Vietnam now sees the United States as a useful counterweight to an expansionist China—Vietnam's traditional enemy for so many centuries.

Ho Chi Minh himself was always suspicious of Chinese intentions, even though they were his allies during his wars against France and America. In fact, Ho once memorably said that if he had to choose between being under the old French Empire or under the Chinese, he'd prefer the French, even while he was battling the French for control of Vietnam. He told his comrades:

> You fools! Don't you realize what it means if the Chinese remain? Don't you remember your history? The last time the Chinese came, they stayed a thousand years. The French are foreigners. They are weak. Colonialism is dying. The white man is finished in Asia. But if the Chinese stay now, they will never go. As for me, I prefer to sniff French shit for five years than eat Chinese shit for the rest of my life.

Now, of course, all of that is history. Mindful of Chinese intentions about territories that Vietnam claims, such as the oil-rich Spratly Islands, Vietnam wants some big friends on its side. Vietnamese government policy is to get closer to the United States, which is why the Museum of American War Crimes in Saigon has been renamed.

Going into what's now the War Remnants Museum, one sees how prominently the officially sanctioned history places early US help in the Vietnamese drive for independence. Exhibits go to great lengths to stress how a US medical team once saved Ho Chi Minh from malaria in the early 1940s, when Ho was leading the early Vietnamese rebels, the "Viet Minh," in their fight against both the invading Japanese Army and the collaborationist Vichy French regime that was then running Vietnam.

The twists and turns rapidly get complicated, the bottom line being that this land and its people, plants, and animals seldom saw true peace for much of the twentieth century.

But as I was to find, all that warfare had paradoxically saved the wildlife. To a certain extent, war had caused the wildlife to remain protected from loss of habitat, safe from hunters or trappers, and undiscovered by the West. No one was going to clear forests full of mines, hunt in a war zone, or go searching for new species near a battlefield. Constant, low-level warfare had given the animals of the "Lost World"—those living away from active firefights but near enough to keep humans away—some breathing room.

It was counterintuitive, paradoxical, and puzzling. I was going to need more time to investigate and confirm, but there it was, in Vietnam and other hot spots, such as the demilitarized zone between the two Koreas.

Of course, no one was arguing that war is good for the environment. Rather, that an accident of history, in the form of Indochina's constant warfare in the twentieth century, gave these creatures a second chance.

But this fluke of history does not mean that anything can be taken for granted when it comes to the protection of these wild animals. Their future is very much in doubt, and their survival in peacetime requires active effort from human beings.

And I was to find that some of the people doing the most to rescue the creatures—and get over the war, move on, and promote reconciliation between the two countries—came from the most unexpected quarters, among people who had survived some of the worst, most unimaginable experiences at one of the most horrific places of the entire American war in Vietnam. A place that even decades later can still bring chills.

3

A Room at the
"Hanoi Hilton"

I still sometimes find myself thinking:
"My God, I'm reporting from the enemy capital."

—David Lamb, *Los Angeles Times* foreign correspondent

IT WAS 1998, and I was walking the streets of Hanoi early one November evening, guidebook in hand, preparing for the upcoming trip to the EPRC.

I discovered I was outside the gates of the notorious Hoa Lo Prison, better known as the Hanoi Hilton—the sardonic nickname given by captured US prisoners of war when they were held here. Because it was originally built by local French authorities to incarcerate any Vietnamese who rebelled against the colonial system, a portion of the structure was kept intact as a historic site, even while the rest was demolished to make way for a new office town.

Appropriately enough, the name Hoa Lo can be translated as "fiery furnace" or "Hell-hole."

It was past closing time, but I knocked on the heavy wood door anyway, to ask about entrance fees and opening hours. A peephole revealed the janitor's face. We went through much gesturing and pointing, then he eyed the street to see if we were alone and said, "Two dollars."

Before I could respond, I heard the unlocking of a series of bolts, chains, and drawbars. After a full minute, he swung open the door, pulled me in, pocketed the cash, held a finger to his lips and motioned to walk quietly down one of the few lit hallways. As I started down the echoing passage, I heard him lock up behind me.

I decided not to think about the legalities of what I had blundered into—was that two dollars an entrance fee or a bribe?—and instead focused on examining the first room's contents: old pictures of the prison, handcuffs, leg-irons, displays, and explanatory plaques in Vietnamese. Dated propaganda photos showed cheery American prisoners next to an English-language placard that claimed, "Though having committed untold crimes on our people, but [sic] American pilots suffered no revenge once they were captured and detained." I made a mental note; this war was undeclared and fought outside the Geneva Convention.

I went through room after room alone while the janitor did his work elsewhere in the huge complex—with the keys still in his pocket. I tried deciphering the different plaques, reading the English portions and puzzling out the rest.

In the last room, a single lightbulb revealed a strange, enormous dark object made of heavy wood timbers. It looked vaguely like a bed frame, with a long, horizontal flat section about thigh-high and eight or ten feet in length. But where the "headboard" should be there was instead a vertical portion about fifteen feet high, with a pulley on top and a tattered piece of old clothesline dangling down. On the adjacent floor lay two stained and well-used containers: one was a round, rusted metal tub about the size of a cheap laundry basket, while the other was a rectangular rattan box about four or five feet long, eighteen inches wide, and eighteen inches deep. Both had sturdy-looking sets of multiple handles. Everything looked worn-out, dented, and fit for the trash can.

No explanatory materials could be found. I noticed a hefty piece of wood at the base of the "headboard" had a semicircular notch about the size of a cantaloupe carved in it.

Then I became aware of ancient, brown-black stains. Suddenly, I realized that they were dried blood.

This was a guillotine.

And not a replica but a working, genuine machine of death, placed as casually as an armchair. There were no velvet ropes or Plexiglas panes between visitor and executioner's tool; no guards; no alarms; no warning signs; no spotlights; no explanatory plaques or audiovisual displays. A single sentence buried deep in a handout said: "This head-cutting machine cut many patriots and revolutionists." This same guillotine must have been the one in place when the American pilots were imprisoned here.

I was as mesmerized as a deer in the headlights. (Apparently, I am not the only one. Nearly ten years later, the *International Herald Tribune* wrote that a French museum exhibit on crime and punishment drew record attendance. The headline said it all: "Guillotine Again Draws Gawking Crowds in Paris.")

Even in the weak light, the guillotine's ominous power was apparent. As my eyes adjusted, I could discern more details, and realized that the metal bucket was for catching the chopped-off head while the long rattan box caught the decapitated body. Judging from the cheap materials and their beat-up condition, there was nothing reverential about the way the authorities disposed of the condemned's remains. The watchwords were ease of use, high turnover, and cheapness. The penny-pinching pettiness implied by this bargain-basement method of disposal gave me a sense of loathing. How can one think so little of the remains of a human being?

But the towering guillotine itself held the allure of power. I couldn't help running my finger in the carved notch of the chopping

block. It was amazing that a tourist in Hanoi could touch such a historic object; everything was more casual in Vietnam.

Unthinkingly, I started treating the guillotine like a Disney
World attraction, even wondering what it would be like to put
my own neck in that half-circle. It held the charm of the forbidden. After all, I wasn't supposed to be here anyway, and how many
people could say that they've literally put their head in a guillotine
and lived?

In for a penny, in for a pound.

Then I noticed that the steel cutting blade of the guillotine was
still in place, ready to go. The more I looked, the more I realized
that it must be the original blade, with little, individual pockmarks
of rust showing in the steel. And in keeping with the museum's quest
for historical accuracy, there was nothing to keep the blade from
dropping: no bolts, braces, supports, barriers, or any other protective
devices. Only a tattered piece of rope fought the force of gravity and
kept that steel blade from sending someone into eternity.

Here I had been treating the guillotine as if it were defanged,
when it was in fact restrained only by a quarter inch of crappy old
manila rope.

And while I am a science writer with no mystical bent at all,
after examining that blade and that rope, there was no doubt in
my mind at all that the damned blade would have fallen had I put
my neck in the carved notch of that chopping block.

I immediately backed away and raced through the halls, looking
for the janitor to unlock the front door and release me.

Outside, I headed for the bright lights of downtown, only ten minutes away. On the way, I passed some elegant young women dressed
in the traditional flowing-white *ao dai,* en route to some formal event
at the newly restored Opera House—an ornate nineteenth-century
building that is an exact duplicate of the Palais Garnier Opera
House in Paris.

The sight reminded me of something I had seen earlier in the countryside, when I had come across a swarm of beautiful white butterflies. Moving closer, I had discovered that the tiny, delicate creatures were feasting on roadkill.

How to reconcile such beauty and grace so close to such horror? What to think of a culture where a special attraction on the tourist trail was a trip to the "Snake Village," in which tourists drank wine mixed with fresh cobra blood while the snake's heart still beat at the bottom of the glass? Where generous hospitality to strangers was combined with the practice of eating the brains of live animals? Where grand buildings were only a ten-minute walk away from the most dismal prison in Indochina?

Later that night, safe in the cheery company of a boisterous crowd from a hostel and drinking home-brewed *bia hoi* beer, I was embarrassed by how much I had frightened myself during my short time in the prison. What would it have been like to be a prisoner of war in a cell there for year after year, knowing that this guillotine was just down the hall, waiting for you? How would it affect your attitudes toward the country, the environment, and the people of Vietnam today?

So, a few days later, I asked someone who would know something about the two extremes of Vietnam: Pete Peterson, former POW at Hoa Lo prison, and now US ambassador to Vietnam—the first high-level American diplomat stationed in the country since the end of the American War on April 29, 1975. This prison and its terrors, both physical and psychological, were what Peterson, John McCain, and all the other US servicemen had to endure—and made Peterson's return to Vietnam as US ambassador, and his embrace of Vietnam and the effort to bring the country back into the world community, all the more extraordinary.

In 1966, Douglas "Pete" Peterson was an American Navy pilot whose F-4 Phantom jet was shot down over North Vietnam. He was subsequently a prisoner of war at the Hanoi Hilton for the next six and a half years before being released.

Eventually, he returned to Hanoi in 1997 to become the first American Ambassador to Vietnam since war's end. Peterson passed the walls of his old cell each day on his way to his office, located just five miles from Hoa Lo Prison. At the time of our conversation, the American embassy was so new that workers were still renovating parts of it, and visitors had to pick their way past dropcloths and open buckets of plaster.

The silver-haired ambassador looked hale and fit, and was dressed like the CEO of an up-and-coming company, in a striped shirt with French cuffs, suspenders, and a yellow power tie. I don't know what I was expecting, but this dapper demeanor certainly didn't fit the bill. To look at him, Peterson didn't seem physically affected by his experience as a prisoner of war, which he once described to an Australian newspaper as "hours and hours of boredom, spliced with moments of stark terror"—although I was to read later that rope burns still scarred his elbows, and one of his hands still went numb from the nerve damage inflicted by the tight manacles.

Our interview took place with his press officer, science adviser, and other staff present. Before we discussed the current situation with Vietnam's wildlife and environment, Peterson touched on the past, explaining that having been a former POW sometimes worked to his advantage when dealing with some of his former adversaries; there was a shared sense of having suffered and endured great privation. "I had seen them at their worst and they had seen me at my worst. . . . I was able to open doors that maybe others couldn't," Peterson declared.

He was probably helped by certain aspects of Vietnamese culture, described by numerous expatriates, visitors, diplomatic staff, newspapers, and guidebooks as "very informal, pragmatic and down-to-earth." From what I could see, the prevailing attitude in Vietnam seemed to be "Now that that's over with, let's get on with the real business." There was not much evidence of lingering over the past or romanticizing the war years; there was no Vietnamese equivalent of what Vietnam War photojournalist Tim Page termed "Namstalgia." The locals were quite willing to establish bars named Apocalypse Now or sell "genuine fake" dog tags to make some money off the tourists, but there seemed to be no interest in that war beyond visiting war graves. It was just one more war fought in a succession of wars in Indochina over the millennia; in the twentieth century alone, the Vietnamese had fought against the French, the Japanese, the Americans, and the Chinese, as well as their Cambodian and Laotian neighbors.

Rather, they wanted to focus on aggressively building the country. The line from the Communist Party leadership was that they wanted to put the war years behind, I had been informed by a young Party member. This desire to move on was echoed by Tran Trong Duyet, the jailer at Hoa Lo Prison who used to guard John McCain. Denying any evidence of mistreatment, he was to tell *USA Today* in 2008, "If I were an American voter, I would vote for Mr. John McCain." The former jailer, seventy-five at the time, expressed an interest in meeting his old captive again. "We would talk about the future, and we would not talk about the past," he asserted.

Likewise, Peterson said he harbored no residual negative feelings. "It's injurious to the United States to ostracize Vietnam; we benefit by engaging the country," Peterson told me. He seemed to genuinely be able to put it all behind him; when I discussed my impression of the ambassador to others, they said that they had the same perception as well—astonishing as it may be.

Afterward, I found that the Vietnamese people I encountered were fascinated by Peterson, particularly in light of the fact that a few months earlier, the ambassador had met a Vietnamese woman with Australian citizenship, Vi Lee, at a Hanoi reception for the diplomatic community and married her. "I think he is very strong in here," was the comment of my English-Vietnamese translator, while thumping his chest and pointing at his heart. Peterson was so popular that ordinary Vietnamese people would stop him on the street and ask to have their picture taken with him.

It was hard to take it all in, particularly after my own encounter with the remains of Hoa Lo. Peterson seemed like an extraordinary individual, more pioneering in his thinking than many people— I don't think it's all that common for a former prisoner of war to go back to the former enemy capital and get married. His attitude of "get over it and move on" tied in to his endorsement of projects between East and West—including projects to protect the country's wildlife and its environment, such as Tilo Nadler's.

"A lot of us have experiences here, but I try to measure the past fleetingly," commented Peterson. "People find it hard to believe, but I've long since reconciled to what happened. As a combatant, you have to accept whatever comes your way. . . . In a situation like that, success is survivability. You win by walking away."

Or, as he later phrased it in a PBS documentary, *Assignment— Hanoi:* "I want to heal the wounds between the United States and Vietnam. It's a tragic history that we've shared as two peoples. No one can change that, but there is a great deal that we can all do about the future."

Accordingly, Peterson told me, "We want to encourage such projects [as Nadler's] between Westerners and the Vietnamese," and pointed out that dozens of veterans' organizations, covering the entire political spectrum, sponsored trips for American vets to go to Vietnam, in which these former soldiers planted trees, made donations to clinics, founded schools, and informally shared information about their experiences during the war and afterward.

More than 230,000 Americans travel to Vietnam per year, according to the latest available Vietnamese government travel statistics. (Only next-door-neighbor China sent more tourists to Vietnam.) A large percentage of these American tourists were war veterans, although their exact numbers were unknown.

Many of these veterans wanted to know more about the land and the people of a place where they had come of age—the age of the average US soldier in Vietnam was nineteen. One, who would only identify himself to me as "Earl, from Pasadena," said, "It was like growing up in a tough neighborhood. Whatever happened there, it was where you became a man." He explained that this was the reason why he had brought his wife with him from the States to see the streets of Ho Chi Minh City. In fact, there were so many American veterans visiting Vietnam that the US consulate had a special section for them on their website.

There was a catch, however. As of our 1998 interview, there were concerns over possible payments for war reparations, and the American and the Vietnamese governments still disagreed over the long-term effects of Agent Orange on the people, biology, and environment of Vietnam, and whether it was responsible for cancers, nerve disorders, and birth defects. Some on the Vietnamese side saw the issue as the basis for asking for potentially billions of dollars in war reparations from the US government. Some private organizations, such as the influential Vietnamese Red Cross, called the spraying of the defoliants a "war crime" and filed suit in US federal court against the private manufacturers of Agent Orange and other defoliants used in Vietnam—the case was eventually dismissed on the grounds that the defoliant's use did not violate international law at the time. Former US servicemen in Vietnam were, understandably, greatly concerned about the potential effects of the chemicals on their health. In the words of the BBC as of 2004: "But since the end of the Vietnam War, Washington has denied any moral or legal responsibility for the toxic legacy said to have been caused by Agent Orange in Vietnam." More scientific studies were called for by the US government, which smacked of delay and stonewalling to some.

In 2006, a book would be published called *Vietnam: A Natural History,* written by three wildlife specialists at the American Museum of Natural History—Eleanor Jane Sterling, Martha Maud

Hurley, and Vietnamese expert Le Duc Minh. In it, one of their studies would say that even in 2006, more than thirty years after the conclusion of the Second Indochina War, scientists were still finding traces of Agent Orange in freshwater animals. The herbicide contains compounds known as dioxins, which are potentially harmful to people and wildlife. The study added: "In addition to effects on individuals, the defoliants undoubtedly modified species distribution patterns through habitat degradation and loss, particularly in wetland systems."

In addition, "Agent Blue"—one of the rainbow of color-coded herbicides sprayed on the country—contained an organic arsenic compound designed to attack the enemy's crops. Repeated applications of the chemicals "sometimes eradicated all vegetation"; according to their study, the environment had still not recovered in many places. Weedy plant species—sometimes native, such as bamboo and rattan, and sometimes "American grass," or *Pennisetum polystachyum*—often invaded these cleared areas, killing other plants and preventing normal regeneration of the forest, much like what I had seen in Cat Tien. In many areas, these weeds continued to dominate the landscape decades after defoliants were sprayed. Because the spray was often concentrated along strategic waterways, it was thought to have had a long-term effect on wetlands and riverside vegetation.

In 2011, the United States was to start a $34 million project to remove dioxins from the airport of the massive old US airbase at Da Nang. In his remarks announcing the endeavor, the new US ambassador to Vietnam, Michael W. Michalak, would say, "The United States and Vietnam have achieved a level of cooperation that would have been unthinkable just a few years ago."

But back in 1998, Pete Peterson explained that these differences of opinion over the effects of the various chemicals had led to some convoluted situations for anyone trying to do any kind of science involving the natural environment. A few days before our

November 1998 interview, an American scientist doing field research on the effects of Agent Orange had been detained at the last minute at Ho Chi Minh Airport and his notes confiscated just as he was leaving the country. Despite the fact that the scientist, Arnold Schecter, had obtained all the necessary clearances and approvals in advance from the authorities, and despite his close collaboration with Vietnamese researchers, all his material had been seized, making months of research in the country for naught. It was apparent that researchers of any kind had to tread a fine line when dealing with the country's environment.

Later, over coffee with a young Party member in an outdoor café across the street from the cathedral in Hanoi, I was told that things were far from monolithic in Vietnam's government, with one department not knowing what another was doing, multiple factions carrying different agendas, and turf wars over sometimes overlapping lines of authority. Apparently, someone felt that the scientist's work might endanger future trade negotiations between Vietnam and the United States.

Despite the ongoing differences over the war's aftereffects, the actual war itself only came up in special circumstances. Peterson's final comment was that most people alive in Vietnam today had nothing to do directly with the war; the majority were born long after it was over. Later, on the *Charlie Rose* show, Peterson was to pithily sum up the situation of the average Vietnamese person on the street in one sentence: "They know as much about that war as the average American high school student—which is nearly zero."

Peterson's comment was certainly true of at least some individuals, such as when I met with a young American primatologist in Vietnam a decade later named Jeremy Phan. Born and raised in California by Vietnamese parents who fled the country at war's end, he felt it was more or less a coincidence that he happened to be doing graduate work in Vietnam. In fact, Phan said that he could not speak any Vietnamese, and had to have a translator whenever he needed to speak to local staff.

Any war-era influences may actually have had more to do with each individual's personal experiences, then. *Los Angeles Times* reporter David Lamb once told me that during the Tet Offensive, he used to watch firefights from the rooftop of the Caravelle Hotel in downtown Saigon. He remarked that he could watch the airstrikes and then go downstairs to his hotel room and type it up for that evening's story. Because of such events, Lamb felt that the war era was indelibly tied to his take on the country, no matter how hard he tried to remain the dispassionate reporter today. Even though Lamb went on to live and work as a foreign correspondent in Hanoi years after the war ended—the only reporter known to have done so—he said, "I still sometimes find myself thinking, 'My God, I'm reporting from the enemy capital.'"

His take on the situation was that the horror and the beauty associated with the country coexisted simultaneously, much like what is found in Mother Nature. Lamb, who was nominated for the Pulitzer eight times for his overseas coverage, and who was to continue as a journalist until he had been in the business for thirty-four years, said that you had to accept that dichotomy as part of the job of reporting on a very foreign culture. And even now, the war and its trickle-down aftereffects were a big part of covering this country—in a book he wrote about his experience, titled *Vietnam, Now,* he wrote: "You cannot write about Vietnam without talking about the war any more than you can write about Saudi Arabia without talking about oil." A thousand years of conflict had shaped this land, with some of the worst of the fighting in the past century. It would take time to accept that Vietnam was a country, not a war.

"But there's something special about Indochina; it gets in the blood and I keep coming back," he said. "It's like malaria."

4

Roughing It in Rural Vietnam

Finding a new genus of mammal is always a shock. That it exists at all shakes our foundation of knowledge.

—John Robinson,
Wildlife Conservation Society, New York

I'D HAD PLENTY of time to mull over everyone's comments while doing the lengthy trek from Hanoi to Nadler's worksite on the edge of the Red River Delta, via a series of increasingly dilapidated motorcycles and buses. As we bounced past the abandoned remains of concrete sentry boxes and other old French colonial–era fortifications of the "De Lattre line," it was obvious I had strayed off what passed for the beaten path in Vietnam.

Just getting this far was an adventure in 1998. At that time, Cuc Phuong was not yet in any guidebook, even though the eighty-six-square-mile national park was the oldest in Vietnam. Making matters worse, the logistics for nearly everything had to be made through a few, government-approved Vietnamese travel agencies, and their emphasis was on getting foreign tourists to a few popular hot spots such as beaches, monuments, or former battlefields like Dien Bien Phu. The idea that someone wanted to go someplace else just drew stares.

Making a phone call to set up an interview was an ordeal in an era when there were no mobile phone networks in Vietnam; you had to bring a hefty sack of coins down to the local telephone exchange office, wait in line to meet with an overseer, tell him the number you were calling and why, and then be directed to a wooden booth to make a call while the operator listened in. Prices were as much as five dollars per minute.

What's more, you could not be certain how freely your subjects could talk to you. A traveler then was very conscious of being in a Communist country, foreign and exotic, with a decided *1984* mentality—on more than one occasion, people asked to meet outside of their offices because they knew that an office mate doubled as a government spy.

Curiously, one Western backpacker told me that as a single woman traveling alone, she actually felt safer, knowing that the government was keeping an eye out for her everywhere. That was certainly an aspect that I had not considered, although the Big Brother angle did work to my benefit once. I had traveled to a wildlife sanctuary unannounced, wondering where I was going to spend the night, only to have someone greet me at the entrance and say, "We've been expecting you. Your room is ready."

Outside of that, difficulties abounded. For foreigners like me, who could not grasp the tonal language of Vietnamese and its regional dialects, negotiating for transportation to a national park or government ministry consisted of pointing to sentences in battered Lonely Planet phrase books and hoping for the best—often with mixed results. (I counted nine different ways to say "hello" in Vietnamese, depending on the recipient's age, gender, social position, education, accomplishments, and the level of formality required. After mistakenly greeting an eminent, prestigious, aged Vietnamese scientist with the version of hello that was supposed to be used for little schoolchildren—during which my official interpreter/government chaperone for the interview turned bright red—I settled for the all-purpose, generic *"Xin chao."*)

Somehow, with the aid of a lot of gesturing at maps, I found myself early one morning at the correct line at the bus station for my three-hour ride. A conductor took me by the elbow and hustled me over the dirt-covered, red marble floor of the terminal to what was thankfully the right bus—a decrepit vehicle with cracked windows, rusted bumpers, and no air-conditioning to fight the already-rising heat. All the seats were taken, but someone was generous enough to squeeze over to let me sit alongside—Westerners were still enough of a novelty to merit special attention. (Many times in my travels, I would stop to take pictures of the scenery, only to find when I put down the camera that a crowd of local people were staring at me with openmouthed wonder. I have to admit that if the situation was reversed, and a complete stranger in traditional Vietnamese garb stood outside my apartment building snapping photos of my front door, I'd be curious too.)

Although I am an average-sized Westerner, I found the seat on the bus so narrow that I could not squeeze in my knees but had to contort my legs into an angle and hold my pack on my lap. Even though the vehicle was already packed, passengers continued to board, and the driver put a narrow wooden plank down the middle of the aisle to act as a bench for latecomers.

I began to understand why Nadler's center had few visitors.

We continued loading. Just when it seemed that no one else could possibly fit, peddlers climbed aboard to walk down the aisles and hawk their wares to the bus passengers—who had to duck to avoid being hit by a wicker basket full of fruit or fresh bread. But the smell of their freshly baked baguettes permeated the air, masking the odors of diesel fumes and sweat.

If one wanted to see and tell the story of the real Vietnam, presumably free from government oversight, this was the price. "The truly authentic experience sometimes requires roughing it," a Western contact had warned earlier, over a get-acquainted lunch in Hanoi.

The bus's occupants made for a cross section of the country: a young man in the brown robes of a Mahayana Buddhist monk was perched next to the door; an older, female Army officer in a starched green uniform sat rigidly erect in front of me; and a shoeless peasant was scrunched up directly behind me—I could tell he was barefoot by the toenails digging into my back. A twenty-two-year-old female student sat to one side of the Army officer and practiced her English on me, asking questions that seemed intrusive to Westerners but are customary in Vietnam: "How much money do you make? How old are you?"

Others apparently listened in and seemed to understand my responses, judging from the occasional nod. At one point I thought she was inviting me for lunch at her house in Ninh Binh (pronounced "NING-Bing"), the provincial capital, but seeing the way that the hairs on the back of the neck of the female officer (her mother?) bristled, I changed the topic.

After we finally arrived at Ninh Binh, there was another surprise: there was no bus connection to the park, more than twenty miles away. After more gesticulating, the owner of a motorcycle agreed to drive me there. We negotiated prices with the aid of a pocket calculator: I would punch in a number, he would clear it and punch in a higher one, and then I'd punch in a figure partway in between, and so forth until we came to a satisfactory amount. He gave me a helmet and I perched behind him, grasping the back of his motorcycle with one hand while using the other to hold on to my knapsack. Clinging to the back of motorcycles was to become a familiar way of travel, known as *xe om*, or literally "bike hug."

Unexpectedly, this open-air mode of travel turned into an

exhilarating introduction to traditional Asia: The breeze created by the motorbike was refreshing, the day was sunny, and we roared past a constantly changing vista of water buffalo, rice paddies, stone cottages, and limestone peaks. We crossed over a bridge with wood sampans tied to its piers, and then past a woman herding a flock of ducklings down the road—causing a squawking whirlwind of feathers to rise up.

The trip was also smoother than expected. Because it was rice-harvesting time, farmers were drying their produce on the hot surface of the asphalt road. This meant that the last few miles of our journey were conducted over a nearly continuous bed of sweet-smelling rice hay, which gave off the odor of a fresh-cut meadow. (The passing vehicle traffic helped trample the rice kernels out. Farmers periodically raked it back and forth, in the process creating a long single strip that resembled the yellow stripe down a median, only one made of rice kernels instead of paint pigment.)

I fell in love with Asia on the spot.

The physical beauty of rural Vietnam was entirely unexpected. Like many foreigners, I had imagined a gray wasteland from all the wars that have been fought over this land. But much of the countryside had been spared the effects of any bombing or defoliation, because the heavy-duty battles tended to be concentrated around key, strategic points as opposed to blanketing the entire landscape. In a country with a landmass about three-quarters the size of California, this meant that places outside the focus of the action were spared—why bother shelling a patch of insignificant jungle in the middle of nowhere?—which partially explained how its wildlife had managed to survive.

But Nadler was another matter. After all that effort, when I arrived at the locked gate to the center, there was no one there.

I killed time by eating a hot bowl of the country's tangy, chili-spiced noodle soup known as *pho* at a store outside the park entrance. It was to be the first of many encounters with *pho;* Hanoi is reputed to make the best version, which includes thin strips of beef. I shared a table with a British entomologist, who told me—with some envy in his voice—that East Germans like Nadler had long held a big advantage over any other biologists who wanted to work in Vietnam. The fact that Nadler happened to be born in what had been a Communist "brother country" gave him at least a five-year head start. East Germany had been a strong supporter of Hanoi during the years of the American War. For years, German—and Russian—were heavily promoted in language schools in Vietnam.

After going back to the gate a few more times, I finally roused someone and convinced him to unlock it and let me in.

Bare-chested in the heat, Nadler turned out to be a deeply tanned and fit man of average height, in his late fifties as of 1998. The hair on the top of his head was starting to thin and turn white, while that at the bottom remained black. The overall effect made him look remarkably like one of his leaf-eating monkeys, or langurs, as his Vietnamese interpreter and live-in girlfriend (later wife), Nguyen Thi Thu Hien, gleefully pointed out.

While the three of us sat outside under the shade of the center's veranda and talked, langurs, gibbons, and other creatures leaped from tree to tree. The gibbons were particularly entertaining. When swinging from branch to branch with their long arms fully extended, they looked huge. But as soon as they stopped and sat down, they shrank into little balls of fur seemingly not much bigger than a very large squirrel.

Nadler himself seemed to curl up inside at first. Like many wildlife biologists I'd met, he initially seemed more comfortable with wild animals than with *Homo sapiens.* He was not hostile, only gruff, sharply warning me away from a quarantine pen. But as soon as we started talking about his favorite subjects, Nadler warmed up, pacing around, hauling out pictures of langurs, taking me to their sleeping "nests," and letting his favorite langur perch on his shoulder for a photo. And sometime much later I discovered that there was a reason behind his zeal in keeping me away from the quarantined animals: because primates are our close evolutionary cousins—considered part of the same order as humans—they are very susceptible to human diseases. As an outsider, Nadler didn't want me to bring in anything contagious. (A zookeeper later confirmed that an effective quarantine station is a hallmark of good animal husbandry.)

Nadler emphasized that the EPRC did not trap animals in the wild and bring them in, but instead rescued creatures that had already been taken from the wild with no hope of return, and whose genes would have otherwise been lost to the population at large. The animals—about one per month—came from diverse sources: many had been confiscated from smugglers by customs authorities, but others were found in marketplaces or had been kept as pets by foreigners before growing too big to handle.

The fact that the animals had already been removed from the wild turned out to be an aspect that other wildlife conservationists immediately seized on when I described the EPRC later. One biologist, George Schaller, commented that once an animal was in captivity, it was out of the loop and no longer contributing to the gene pool, unless extraordinary measures were taken to shuttle animals or their DNA back and forth—an expensive and not always successful undertaking. Comparing the situation of langurs to that of other species he had worked with, Schaller said, "There are about eighteen thousand tigers in captivity. Those are essentially dead tigers, for all the genetic contribution they make to the population; they're not going to be reintroduced to the wild."

Another researcher, Joe Walston, then the Cambodia representative of the Wildlife Conservation Society, commented, "I would certainly be less of a fan of Tilo if he removed remaining populations from the wild, but the fact that most of these animals came from confiscations implies that the wild population is not being directly threatened by his work."

In addition, because the EPRC was dealing with animals so extraordinarily rare, any birth—whether in captivity or in the wild—was considered a notable success for that species; there were no other conservation efforts for the langurs. Sadly, for some animals, the EPRC may be their last home—but the EPRC could at least keep them alive and well, in a comfortable environment, thereby offering a healthy reservoir of wildlife, if and when a safe place could be found in the wilderness where the animals could be released.

Surprisingly, it turned out that many zoologists have mixed feelings about keeping wild animals in confinement and even the very idea of captive breeding; this includes zoologists affiliated with some of the biggest and most prestigious zoos in the world. This was a completely unexpected development, akin to a NASA astronaut questioning the space program. Nevertheless, Alan Rabinowitz, who is on the board of the Bronx Zoo, said, "Zoos often become jails for animals. I'm not a fan of zoos for zoos' sake, and the argument that they are 'educating the public' can be overrated, depending on the zoo and the situation." However, Rabinowitz added that there were exceptions: "Admittedly, there are some instances where a zoo is a really good alternative, such as if the animals are injured, with a broken wing or busted leg, or if the animals would be otherwise extinct."

Critics charge that zoos are an artificial environment, with carefully constructed props—such as "termite mounds" filled with jelly—installed "as much for the entertainment of the humans gazing into the exhibits as for the animals themselves," wrote Thomas French in his book *Zoo Story: Life in the Garden of Captives*. "Man-made waterfalls and other special effects, such as fake rocks made to appear weathered with air-brushed mineral stains and simulated bird droppings, encourage zoo visitors to feel as though they are witnessing wild creatures in their natural habitat . . . the principle guiding such aesthetic touches has been called 'imitation freedom.' Animals of any intelligence, presumably, are well aware of the difference."

Further criticism was voiced by a correspondent in Southeast Asia, Nate Thayer, who claimed that plenty of non-governmental organizations (NGOs) are in the "save the wildlife" racket in Cambodia. NGOs are rife around the planet, and a certain percentage are "fund-raising rackets," including roadside zoos and "farms" that sell animal body parts to the black market. "Certain things attract donations, like tigers, elephants, and the kouprey," Thayer said. "But they [the people running them] too often don't actually leave their air-conditioned offices to actually go to where these beasts have fled to. Because it is not fun; it is dangerous and a pain in the ass in general—which is why the beasts still survive in these places. . . . One of the biggest threats to endangered species is 'wildlife preservation'—ecotourism—focusing on where these animals live. They become cash machines."

Thayer added that there were many committed scientists and many genuinely concerned international organizations devoted to wildlife protection, so he didn't want to tar them all with the same brush. "But the way the international funding for this works is a racket," Thayer claimed. "You can see it in Africa, too. Where 'wildlife officials' are in cahoots with poachers. I am not saying that it is all or even most, just enough to be annoying and dangerous. That is why we never released the exact coordinates where we knew the kouprey was—at a very small radius of watering holes in the peak of the dry season."

Similarly, captive breeding has its detractors. Walston, for one, thinks that captive breeding can detract from the main issue—conserving wild populations of species and their habitats. "The disparate philosophies of conservation and animal welfare are often not compatible and share few similarities, though they are often confused in the minds of the public," he said. "However, Tilo and the EPRC (they are hardly divisible) are an example of captive breeding playing a key function in conservation of some species, if not their habitats. In instances where species are not being protected in the wild and their status is critical, then often captive breeding is an alternative that has to be considered."

Or, as Rabinowitz reluctantly put it, "When the number of a species is down to the last few digits, you don't have many choices. In those circumstances—it's not black and white—you do have to breed in captivity."

The consensus seemed to be that captive breeding and keeping an endangered species in confinement—or even holding samples of its genes on ice—did not necessarily have to be a bad thing, as long as these practices did not encourage people to think that the species was now any less threatened. Their argument was that a species is not just defined by its appearance and color, but by its habitat, behavior, and its relationship with other organisms. If you failed to conserve wild areas and the species for which that area is their habitat, you would lose systems, interactions, and processes that could never be re-created.

Walston may have put it best: "Tilo and the EPRC are working at the critical end of the scale, potentially at the end of a species' existence, where little is being done to protect their habitat. Here, captive breeding, in the way that it is being undertaken by Tilo, is currently their major hope. However, if Tilo manages to both secure breeding populations *and* raise the profile of the species themselves, then maybe original habitats may at last receive the level of protection they deserve."

When I asked when was the best time to see the animals roaming around the treetops, Hien said late in the morning.

"How late?" I asked, with visions of sleeping in after the day-long journey from Hanoi.

"Six o'clock," she replied.

But the creatures' antics the next morning, and their distinctive, plaintive calls, made it worthwhile.

After feeding the animals, Nadler and his staff fed themselves a breakfast of German sausage, cheese, and tropical fruit on the center's veranda. They conducted a little staff strategy session about animal husbandry while they ate; an important side benefit to the EPRC experience has been that zoologists now know much more about the required food supplies and environmental and health conditions that the animals need to survive in confinement—and, presumably, what kind of conditions are needed for them to survive in the wild. Among those seated were Hien and Roland Mannel, a zookeeper who divided his time between here and Germany's Leipzig Zoo.

They chatted in a mixture of English, German, and a sprinkle of Vietnamese, going over the health of the latest animals dropped off on the EPRC doorstep. The conversation centered on a newly arrived langur that looked like it may not survive much longer. They kept mentioning something called *kut,* which I could not find in my Vietnamese phrase book. When asked what it meant, there was a pause, and then laughter when Hien explained that the word refers to animal dung.

One of the best indicators of an animal's health is the quality of its stools; ultimately, *kut* can help tell staffers whether they are feeding each species of langur the right varieties of leaves. Getting the proper diet to all the different primates was a major challenge, and the center's few helpers and seven full-time leaf cutters needed to collect more than two hundred pounds of fresh leaves from the jungle every day. It could be hard to tell if they'd gotten the right leaves to the right animals, especially when dealing with new arrivals, which sometimes showed up on their doorstep as though the EPRC were some kind of orphanage. Generally, the first week was the most critical; if the animal survived two weeks, then it was likely to live for a long time. A lot of finger crossing was involved;

the EPRC once had to guess at the diet of a "black langur," a species previously known only from a single skin collected in China in 1924.

As a result, the breakfast table was the site of many earthy, informal, and pragmatic discussions, much like the down-to-earth culture of rural Vietnam. I wondered if his personality, combined with his familiarity with the Communist system as a native of the former East Germany, helped to explain Nadler's rapport with the Communist Party officials who oversaw Vietnam's environmental-protection efforts.

Breakfast finished, Nadler conducted a tour of the facility, including its large, shaded, airy pens full of native plants and primates, as well as its clinic and care center. Overall, he gave the impression that he would much rather be in the bush, instead of attending to administrative details or public relations.

But Nadler opened up later when he discussed his indirect route to primate biology. "Biology was possible to study only if your parents belonged to the working class and they were members of the Party," he said. To pursue his interests, he had to learn how to work the system. He earned a degree as an air-conditioning engineer but studied biology in his free time, eventually wangling a job at a biological research institute, doing fieldwork, and writing more than fifty scientific papers, a book, and a monograph on birds. Nadler accompanied organized expeditions to Antarctica, Mongolia, China, Russia, and the tiny island of São Tomé, off the west coast of Central Africa.

In 1991, while visiting Vietnam, he saw a pair of Delacour's langurs for sale in a meat market and found himself determined to do something for the creatures. Asked why, he gave that same dismissive look that climbers give when asked why they feel compelled to ascend the north face of the Eiger in Switzerland—if you have to ask, you'll never understand.

Instead, Nadler wordlessly pointed to a photo of a family of

Delacour's langurs at his rescue center, in which a baby rested in the arms of its mother while its father stood behind her shoulder. The photo resembled a studio portrait of a typical human family, fit for placing on top of any office desk. All three animals gazed at the camera, their bright and expressive eyes making contact with the viewer. The bright orange baby fur of the young one contrasted dramatically with the deep, rich black-and-white fur of its mother and father. Despite eons of evolutionary distance between us, it was almost possible to read humanlike emotions into the facial expressions of the young parents with their child.

"Doing nothing would have been worse." Nadler shrugged.

With two Delacour's langurs confiscated from a meat market, Nadler scoured the literature on how to care for them. He didn't know it then, but the EPRC was on its way, and he eventually convinced the prestigious Frankfurt Zoological Society (FZS) and other professional zoological organizations to get involved and give advice. The center got help early on with issues of park management and protection from the FZS, which was the very first of what eventually became dozens of wildlife organizations to operate in Vietnam. With their assistance, Nadler was also able to set up a veterinary clinic and a quarantine station. "It was not my work, not my profession, but what else could I do?" he asked. Today the Endangered Primate Rescue Center has an excellent reputation and is recognized as one of the best-managed stations in Southeast Asia.

Located between village and park, the center acted as a temporary holding center, or ark, for endangered animals that had nowhere else to go. Before coming here, most of the creatures had been destined as trophies for collectors, for use in traditional medicines in China or as delicacies in the specialty restaurants that have sprung up in Vietnam in recent years.

Prior to the EPRC's existence, these animals had faced uncertain prospects, even after authorities rescued them from the black market. The langurs in particular had a tough time. These long-tailed primates—which range in size from seven to forty-six pounds, depending on the species—preferred the leaves, seeds, and unripe fruit of trees that grow in limestone-rich environments, and they have specialized salivary glands and stomachs containing bacteria able to break down only these materials. If fed the wrong thing, such as ripe fruit, they quickly become ill and die. Consequently, Vietnam's zoos did not have the ability to care for them, and few survived shipment outside the country. Most confiscated langurs were simply dumped into the nearest park, regardless of their diet or place of origin.

Nadler's facility provided an alternative, nursing rare and endangered primates back to health in a small patch of protected natural environment free of poachers. From a single cage and a pair of langurs, the center had grown to several dozen cages and a fenced five-acre outdoor enclosure. The animals get to spend much more time in the outdoors here at the EPRC, where the conditions are as close to natural as possible, than they would at a comparable facility in Europe or the United States, where the seasons and the climate are different. Nadler estimated that in Europe, for example, the langurs would only get a couple of months outdoors, while here, they can spend almost all their time outside, eating familiar foods in a familiar environment and keeping their social structures.

About three zoological institutions now provided guidance and some financial help, but a large part of the EPRC's aid came in the form of donated labor from visiting volunteers. Nadler had a gift for enlisting the help of outsiders: his first animal keeper was a nineteen-year-old girl, Manuela Kloden, who responded to an ad at the European zoo where she worked and wound up staying in Vietnam for two years. An electric-fence company donated the enclosure that keeps the gibbons in and the poachers out. And a European electrician came to Vietnam at his own expense to spend three weeks installing it for free.

Defying the odds (some langurs suffered such severe injuries to

their tails that these appendages had to be cut off) the center managed to keep thirteen rare species alive. Five species that called this
center home were found nowhere else in captivity. And Nadler's
center recorded the world's first captive births of Delacour's and Ha
Tinh langurs.

"Tilo is doing something unique," wildlife mammalogist Colin
Groves had told me. He said that the conservation situation in Vietnam was so critical that these langurs "might not survive at all but
for Nadler's efforts."

For the moment, Nadler and his staff have not been able to
think about anything more ambitious than just keeping the endangered animals alive, as a sort of emergency reserve. It was a
familiar situation to conservationists all over the world, whether
they were working with jaguars in Central America or antelope in
Tibet. And last-ditch rescue efforts were not confined to animals,
but plants as well. In the Hawaiian Islands, "extinction capital of
the world," researchers rappelled down cliffs in order to find and
rescue native plant species at risk of extinction for possible breeding in captivity—with similar issues about their strategy. Known
as "extreme botanists," they risked their lives to bring plants back
to the National Tropical Botanical Garden—located, appropriately
enough, on the "Garden Island" of Kauai. It was all part of a de-
centralized, grassroots, worldwide trend to save endangered species
across the tropics, home to most of the world's biodiversity.

(Since 1994, the global conservation status of a given species is
determined by groups of specialists who evaluate each species using
information on population size, rates of decline, and degree of habitat
fragmentation. The World Conservation Union publishes this "Red
List" of their results, with species placed into a number of different
threat categories, ranging from Critically Endangered to Extinct,
including special categories such as Extinct in the Wild—meaning
that a few stragglers may still hang on in zoos.)

Nadler was considering moving things up a notch at the EPRC, and seeing if they could try to return creatures to the wild. But it promised to be a tricky business, as it can be hard to match the right animals with the right habitat. In addition, they didn't want to cause conflicts with members of the same species already living there, or outrun the local food supply. There was also the problem of releasing animals into the wild that had been in captivity too long—it would be something like throwing a poodle out to live among the wild dogs of Africa.

But they would have to try something soon, because focusing solely on rescuing and breeding captive animals was not an answer either. "Technically, you could keep a species alive through saving genetic material or even captive breeding, but in reality this is nothing more than operating a life-support machine for a body that can no longer operate independently," said Joe Walston.

While he said Nadler's center is "the major hope" for Vietnam's rare primates, Walston pointed out that conservation should be about more. "It is the complex landscapes that we should be protecting, not merely identifying and bottling a few of their constituent elements. Conservation should be about maintaining the integrity of natural systems, often called 'ecosystems,' and those species that exist within that system. These landscapes are not static like captive species, but constantly competing, adapting, and changing—sometimes resulting in natural extinction, sometimes facilitating the creation of new species. It is these complex landscapes that we should be protecting, not identifying and bottling a few of the constituent elements."

Nadler agreed but felt that time was running out. He did not see temporary holding facilities as a rival to habitat conservation, but an adjunct. "The biggest problem in Vietnam is that there is no time for education on environmental issues. It takes twenty years to see the effects of an education program, and these species don't have even ten years," he said. "You can't put animals back in the wild if they are extinct."

To try to address the larger conservation issues, Nadler and his staff worked closely with Dao Van Khuong, then director of Cuc Phuong Park. Dao took a multipronged approach to wildlife protection, one aspect of which was to provide local people with alternatives to cutting down jungles or hunting endangered species for their livelihoods. With Dao's help, a deer farm had opened on the park's outskirts, where the horns of *Cervus nippon* were harvested for use in medicine. Officials were also promoting sustainable alternatives such as beekeeping and the raising of civets, whose body parts were in great demand by the perfume industry.

Dao said that the biggest problem the park faces was the burgeoning human population, which was consuming all the resources needed by the wildlife. It was a familiar story in developing countries, at a time when countries such as India were seeing their human populations increase by 58,000 per day. In 1962, when Ho Chi Minh founded Cuc Phuong National Park, there were only 500 people living nearby, and the only visitors were scientists and government officials. Dao told me that when he became a ranger here in the early '70s, there were still tigers around. As of a 1998 survey, there were 51,000 people living in the region, and 60,000 visitors. The result was that deforestation was creeping into the park.

And there were no tigers anymore.

Nadler and his Center couldn't do anything about Vietnam's population growth, especially on their budget of less than $20,000 a year. But one thing they could do was create a climate more conducive to conservation. By linking zookeepers and biologists from the West with government officials and citizens in this formerly isolated Communist state, the EPRC helped to build skills and give the Vietnamese a sense of pride in their native wildlife.

Would these efforts be enough to halt the slide of rare primates such as the Delacour's langur into extinction? The odds were clearly against their survival. But Nadler believed it was important to at least make an attempt. "At the rate that the Delacour's langur is disappearing, for example, it is unlikely that future generations will ever know this species outside of history books," Nadler wrote in a Conservation International newsletter. "Still, we have a window of opportunity to save them."

And, thanks in part to the efforts of the former East German engineer and his staff, the *kut* had not yet hit the fan.

2010

and Beyond

5

A Grand Tour of the EPRC

Hope is important because it can make the present moment less
difficult to bear. If we believe that tomorrow will be better, we can
bear a hardship today.

—Thich Nhat Hanh, Vietnamese monk,
peace activist, and writer, barred from Vietnam
from the fall of Saigon in 1975 until 2005

WHEN I HAD last seen him face-to-face more than a decade ago at
his field station in Cuc Phuong, Tilo Nadler had been posing for a
photo with his favorite langur perched on his shoulder. This time,
Nadler looked a little thinner and a little grayer, but more sprightly
and outgoing than I remembered, and he smiled a lot more. Over
the next few days, I was to get the feeling that much like creating a
start-up company from scratch, the first few years were the hardest,
and now that the EPRC was more established, he had a chance to
enjoy the process.

"This is probably the only place in the world where you can
microwave a cup of coffee and watch Delacour's langurs from the
porch," he says with a laugh.

Tilo was still clad in sandals all the time, but he now wore a uniform with an EPRC logo as well—in fact, I never saw him take it off. He seemed to always carry a camera and several lenses, but never carries a weapon. Tilo's only concession to age was that he dons glasses on occasion; otherwise, he seems little changed, and very vigorous.

He is also still abrupt and no-nonsense: once, when a stereo was making too much noise at a late-night party of wildlife rangers, Tilo marched up with a pair of scissors and cut the cord to the loudspeaker. All he said was "It was disturbing the langurs."

His old girlfriend, Hien, was now his wife, and they were the proud parents of two young children, who scrambled about the office—a place adorned with antlers, skulls, and local artifacts, including a crossbow hanging on the wall. One child has the traditional-sounding Vietnamese name of Khiem, while the other carries the decidedly Teutonic name Heinrich. The two are always making crayon drawings of animals (at least, I think that's what they are) when not practicing their Vietnamese or their German language lessons.

For his part, Tilo still has a pronounced German accent to his English—the language that seems to be the common denominator that everyone lapses into by default at the EPRC. I can pick up the occasional oddity of word choice now and then in his speech, such as when he calls the computer "he" or "him."

Hien is less of the rebel that I remembered; she used to proudly say, "I am not a typical Vietnamese girl." But she still loves to organize and run things, fussing over Tilo and the center like an indulgent mother hen. (I also notice that her name is the only one on the "Contact" section of the EPRC website.) When I meet her this time, she clasps her hands and bows her head in a Buddhist-style greeting—something that daughters of government officials never did in a country where religion was the opiate of the masses.

But what about his work? Where did things stand now for

Vietnam's wildlife and Vietnam's jungles, just before the UN's International Year of Forests? The last time I saw him in person, he had said that the langurs didn't even have ten years left unless something was done; it was now a dozen years later.

It was apparent that many things were different in Vietnam. Even before my plane landed, I noticed changes since my very first visit. The Hanoi airport itself was refurbished, with a new annex containing modern stores where you could buy SIM cards to operate your mobile phone and make local calls at reduced rates—although service proved to be nil outside of the city and its immediate area.

As for the capital city itself, where there once were hordes of bicycles crowding the streets of Hanoi, there were now hordes of motorcycles; where there used to be the occasional rare motorcycle, there were now automobiles. Streetlights no longer went off at ten p.m. every night. The first fast-food stand, a KFC, had inserted itself into the main street leading from the airport to the downtown of the former "Paris of the Orient"; *The Economist* says that the corporate parent of Kentucky Fried Chicken is one of the most aggressive Western multinationals penetrating the Asian continent. A near-permanent gritty concrete dust permeated the air, where office towers now sprung up like bamboo after a rain. Ironically, about a twenty-minute walk away from the old Hanoi Hilton prison, there is now a bona-fide "Hilton Hanoi Opera Hotel," carefully named to avoid any uncomfortable associations.

The presence of the Communist Party was also more discreet and sophisticated these days, even if it was still actively manipulating things behind the scenes. Reporters Without Borders labels Vietnam an "Internet enemy," citing Vietnam's tight grip on information; the government slaps fines in the thousands of dollars for publishing anything deemed "not in the interests of the people." While I was there, the BBC website was down for three weeks, Facebook and Twitter would mysteriously become inaccessible, and phone numbers of key contacts would sometimes stop working overnight. All of which were reasons why I had felt I was only getting part of the story on the status of Vietnam's wildlife and environment from afar, despite numerous long-distance interviews and follow-up stories I'd done from back home in the States.

At the same time, however, a young, low-level contact in the Vietnamese government told me that he had "become addicted to the Internet."

There was a perpetual cat-and-mouse game going on with the censors, which I saw one night at a youth hostel in Hanoi. Equipped with free wireless, it was always full of young people with laptops, who would joyfully share information about how to outfox the authorities and find holes in the country's firewall. One cried out "Facebook is open if you log in at their portal in Denmark" over the noise of happy hour one night, and there was a rush from the hostel's bar as people abandoned their free home-brewed *bia hoi* for a chance at unrestricted Internet access.

Transportation had changed as well. Modern, fully air-conditioned vans, equipped with on-board DVD screens, now take people from downtown to distant national parks and tourist attractions. I no longer needed to jounce along in a dilapidated local village bus and feel a farmer's toenails digging into my back with each bump. Neither did hawkers at each pit stop sell chunks of raw sugarcane to be sucked on like candy. But for old times' sake I chose to complete the final fifteen minutes of my trip on the back of a motorcycle, doing the good old "bike hug."

I had not been sure what to expect on my return, and was gratified to see that the EPRC had not only survived, but thrived. I had been kept apprised of the situation from afar via all the modern tools, but there was no replacement for "ground-truthing."

Reports had varied greatly as to the country's wildlife situation. For every positive story of a new species discovered, there were an equal number of stories about an environment crushed and an ecosystem fast disappearing, due to habitat loss and rapid industrial development. Miraculously, Indochina had continued its wild-life discovery streak, with more new species popping up—since 1992, naturalists have found sixty-three new vertebrates, forty-five unknown fish, and three new deer species. Many are "endemic," meaning that they are found only here and nowhere else. Even some of the best-known species still look odd to Western eyes, such as the Annam flying frog (*Rhacophorus annamensis*), which glides among trees with its webbed feet; the fishing cat (*Prionailurus viverrinus*), which loves to swim when it is not sitting on river-banks and pouncing on fish; the finless porpoise (*Neophocaena phocaenoides*), which lacks a dorsal fin; and the raccoon dog (*Nyctereutes procyonoides*), the only canid known to hibernate. Even the names are extraordinary, such as the beautiful nuthatch (*Sitta formosa*) and the golden Kaiser-i-hind (*Teinopalpus aureus*, a type of butterfly).

But in the past decade, Vietnam had also become an "Asian tiger" of business, with mixed consequences for its wild denizens. Vietnam's economy has been growing by an average rate of 7 per-cent a year for the past ten years, at a time when the economies of mature countries such as the United States and Germany are lucky to grow about 3 percent annually. Like many tropical countries, Viet-nam has a young and rapidly growing population, which expanded by nearly one-third in that time period. (In the region around Cuc Phuong National Park, the average family has 6.7 children. One particular couple has 15.) Economists and financial pundits say that, due partly to its young workforce and partly to a society-wide eco-nomic restructuring in the mid-'90s known as *Doi moi,* or "re-newal," Vietnam can be considered to be "the next China" in terms of economic growth for the foreseeable future. Or, as Alan Rabin-owitz called it, "a miniature China on amphetamines."

I had gone to see for myself—this time, purely on a tourist visa, in hopes of having freer access that might enable me to slide under the radar.

. . .

After we catch up on events—the old park administrator had retired, the deer farm is gone, but a new turtle center was established—we sit down in the front office of the EPRC, in almost the exact same spot as years ago (although the office is now larger and better equipped, and there are many new companion buildings as well, in addition to a house on the premises for the Nadlers). Tilo tells me that he is guardedly optimistic about the fate of some specific wildlife, if worried about the overall shape of things nationwide. "It's all a long way from the early days, when the EPRC was just starting out," he said.

Back in 1991, when he first arrived as a short-term visitor from the former East Germany, he'd had to deal with hazards all too familiar to travelers in these parts, such as the all-pervading wet of monsoon season, which lasts from mid-November through March. In this damp and warm environment, everything rots or breaks down spectacularly fast. It's an ideal situation for termites, which are so aggressive that even the telephone poles have to be made out of concrete or else they get ground down to nubs in just a couple of years, as if they were in a giant pencil sharpener.

Meanwhile, the local flora grows very fast. "I found fungus and mold growing inside the lenses of my binoculars," he says, holding up a pair for me to see.

There were also the problems of dealing with unfamiliar foods, antimalaria medications, and dangerous creatures. In a very blasé tone of voice, as if it were just another day at the office, Nadler describes how he once extracted a king cobra (*Ophiophagus hannah*) from under the kitchen sink in his living quarters. Another time, he stepped outside his front door onto a thirty-centimeter-long (twelve-inch) poisonous centipede with his bare feet. It promptly bit him, and Nadler had to be rushed to the hospital in tremendous pain, which he told the doctor ranked "about a nine-point-five on a ten-point scale." For a long time, he had two little holes in his foot as a memento of the encounter. He grimaced at the thought while glancing down at his sandals—which I'd noticed that he always seems to wear, no matter what the occasion.

Now I know why.

In those early days, Nadler had little formal information on how to raise langurs—not that much existed—and there were not many amenities: no groceries available at the park, no showers, no heat in the wintertime, no phones, and often no electricity. The road to the park entrance was dirt, and there were no computers, copiers, or fax machines. To send an e-mail, he had to do that bus ride to Hanoi and back.

At the same time, he admits that there were some benefits to being so isolated: there were few distractions and few visitors. Nadler says he used to be able to sit down and have coffee with every Westerner who made the journey to his rescue center, or an average of once per week.

There are now about fifty Western visitors to his EPRC per week—as well as phones, electricity, heat, running water, and several computers. And Cuc Phuong National Park itself now gets about eighty thousand tourists per year, of which about 10 percent are Westerners and the rest Vietnamese citizens, usually in busloads of about fifty people at a time. The main visitor center for the park is located just a quarter mile from the EPRC's front gate, and the big tour buses rumble by constantly on the paved asphalt road on the weekends, when thousands of people descend. On Mondays, there's rubbish strewn everywhere on the visitor-center parking lot, a concrete expanse the size of a football field.

Back in those early days, as the founder of the very first—and for many years, the only—wildlife rescue center in Vietnam, Nadler had plenty of time to himself, even if he also had to do everything himself as well. This extended to writing a 264-page illustrated field guide for the rangers—a three-year project—which included such basics as a chapter on animal tracking. In addition, he conducted a series of surveys on the number and type of wild animal species living in northern Vietnam's jungles, and kept a lookout for endangered animals illegally sold in meat markets.

In 1993, the first year of formal, full-scale operations at the EPRC, his animal rescue center started with two China-bound Delacour's langurs that had been confiscated from the local market in Nho Quan. He and his helpers quickly learned that they could not save everything, so he decided that the EPRC should focus purely on rescuing the rarest of the rare: langurs—which were difficult for most full-fledged zoos to keep because of their specialized diet—and the fruit-eating "lesser apes" known as gibbons. "Ideally, conservation should focus on one special species or one special place. We cannot save everything everywhere; instead we should put our efforts into a few key animals in a few key hot spots," Nadler declared.

Since my last visit, the center's work had become more sophisticated, expanding to include 150 animals from fifteen species and subspecies. At the time of my first visit, it had 55 animals from thirteen species. Its track record for babies then totaled seven; it now had more than one hundred. (In some ways, the EPRC had some natural advantages: it dealt with relatively small animals that don't take up much space, compared to large animals such as rhinos, which need large territories to sustain themselves. In addition, while gestation is seven months for a langur, it is close to a year and a half for a rhino, depending on the species. The starting numbers of animals available were different too—in the case of Vietnam's Javan rhinos, biologists were down to the single digits, while langurs could still be counted in the dozens or low hundreds. Too close for comfort, but enough to tilt the playing field in favor of langurs and against rhinos—which were pretty much at extreme opposite ends of the scale.)

Of the animals at the EPRC, six of these primates are found

only here and nowhere else in captivity in the world. In addition to the Delacour's langur, they include the Cat Ba langur *Trachypithecus poliocephalus,* the gray-shanked douc langur *Pygathrix cinereus,* the Laos langur *Trachypithecus francoisi laotum,* the black langur *Trachypithecus francoisi ebenum,* and the Ha Tinh langur *Trachypithecus francoisi hatinhiensis.* Each is found only in small, specific parts of the country; much like Darwin's finches, each species has evolved from a common ancestor to become a specialist that feeds in its own very particular niche.

Because of their extreme specialization, the act of keeping all these different species alive is extremely difficult to do in a zoo: the douc langur, for example, comes from the sunnier, warmer parts of southern Vietnam, and cannot withstand the colder temperatures of the north, where the EPRC is located. So, Tilo and his crew found by trial and error that the doucs' pens needed heating units during the winter.

What's more, none of the langurs travels long distances well, especially if going to zoos in Europe or North America. (There are few adequately staffed and properly financed zoos in Southeast Asia, Nadler says.) Part of the reason may be the climate at their destination: at a zoo in, say, America, langurs can only get a maximum of two months outside, which contributes to their tendency to fall ill.

Some places, like California's world-renowned San Diego Zoo, have gone to great lengths to try to keep their langurs alive, attempting to mimic what happens in nature thousands of miles away in Vietnam. The San Diego Zoo is known for its realistic environmental settings and its lack of traditional cages; its zookeepers created a specially designed environmental "enclosure" for their collection of douc langurs (genus *Pygathrix*)—a somewhat more common species of Indochina langur, described by at least one biology handbook as "one of the world's most beautiful primates." It is fitted with man-made trees, concrete branches textured and painted to resemble bark, detachable vines that can be clipped to branches at different heights, solar blinds to control the amount of sunlight, heated floors, and advanced climate controls to duplicate the effects of monsoon season.

These efforts don't stop there. The San Diego Zoo also feeds its douc (pronounced "duke") langurs a specially formulated biscuit containing vitamins; minerals; vegetables such as kale, broccoli, and dandelion; and leaves such as ficus, eugenia, mulberry, and willow. The San Diego Zoo's newsletter, *ZooNooz,* said that out of the eight hundred species, or four thousand individual animals that the zoo has under its care: "They [douc langurs] are considered among the top 10 most finicky eaters at the Zoo." As late as 1999, its zookeepers reported that "caring for douc langurs in captivity has proven to be an undertaking of trial and error," citing "a paucity of information" and saying that "keepers know little of their dietary or social behavior, and have had measured success in duplicating either."

In contrast, it is much easier keeping langurs at the EPRC—even the very rare and unusual ones. The center is purposely located half in and half out of the surrounding jungle, where the climate, sunlight, surroundings, and temperature are more compatible with the needs of these demanding creatures. It's more a matter of leaving things as natural as possible while keeping the animals safe from poachers. (Of all the langurs, the douc langurs may be particularly susceptible to hunting; when disturbed, their instinct is to respond by remaining motionless rather than fleeing. According to at least one source book, "A ship's crew putting in at Danang in 1819 was able to shoot more than 100 Red-Shanked Doucs between five in the morning and breakfast.")

Probably the biggest change was that the EPRC was now sending animals to other places—and even releasing a few back into the wild, which I was to learn more about. They've started out slowly, releasing a few radio-collared lorises a few years ago, then carefully moving on to the rarer, larger animals.

With some species, the EPRC really had captive breeding

knocked—there were now thirty-eight Ha Tinh babies born at the EPRC, making it seem almost an everyday occurrence, in comparison to the excitement of the very first Ha Tinh baby. The center has done so well, in fact, that the EPRC has now started to send some animals to other rescue centers.

Since my last visit, the EPRC also created a sort of halfway house for primates under consideration for release, in addition to the old five-acre semiwild grounds. This new "training facility" consists of a roughly eight-acre hill of primary forest carved out of Cuc Phuong National Park and adjacent to the cages/enclosures of the center's main area. Surrounded by an electric fence, this zone is where you can hear gibbons hooting and crashing through the trees. It provides a wonderful opportunity to learn about these creatures; much about gibbons, who live high up in the canopy, is still unknown. In particular, their calls have been the subject of much interest.

Gibbon pairs sometimes sing in unison, starting their duets from shortly before dawn. Some scientists suspect that their song serves as a way of marking territory, while others like to think it might be a precursor to the music that we humans—fellow primates— make. Their calls certainly are inspiring, and some biologists have even used the word "operatic" to describe them. Martha Hurley, co- author of *Vietnam: A Natural History,* wrote: "The female produces the 'great call,' an ascending, somewhat eerie, yet high-spirited whistling call that the male accompanies with more staccato vo- calizations, usually adding a coda after the female finishes. Heard for more than a mile, these calls advertise the pair's presence, cur- tailing trespassing and reducing confrontations with other gibbon groups."

"Eerie" and "high-spirited" certainly describe the way these calls felt to me in the half light of predawn, as the sounds wafted past the gauzy cloud of antimalarial mosquito netting surrounding the bed. I never needed an alarm clock while I was there.

Giving me a grand tour of the grounds, Nadler is as proud as a doctor in a maternity ward, rattling off the number of new babies of each species, their mothers' health, and their survival rates. Each cage has a metal label on the outside, affixed with the occupants' scientific name, sex, and acquisition date or birth date. Every animal is assigned a number by species, so there is Cat Ba #15 or Ha Tinh #2.

Though it is decidedly not scientific, staff members also find themselves giving pet names to some individual animals as well. Some of the creatures do seem to have distinct personalities. One favorite is a red-shanked douc named Boots, whom they've had for more than ten years now. Boots will actually let Tilo hold him and groom his soft, luxuriant fur. "He's a real friendly guy," comments Nadler while petting him.

Nadler adds that lately, the market for live langurs has centered on babies, which buyers find small and cute. The problem is that the adult females can be very resistant to anyone taking away their young ones, and the poachers can't always tell which female has a baby suitable for the black market. The result? "Poachers have to kill ten mothers in order to bring one baby to market," said Nadler.

Of course, there's more to it than the plight of an individual animal. It is in precisely such circumstances that the most dramatic struggles are being fought to protect the world's biodiversity.

The greatest mass extinction in our planet's history since the age of the dinosaurs is happening here and now, with most of the action located in tropical, rapidly developing Third World countries such as Vietnam. How this one country deals with its ecological problems, and saves or does not save its wonderful menagerie of newly discovered animals—eight of the ten large mammals found in the past two decades were discovered in Vietnam—could be a foretaste of what happens elsewhere. What Vietnam does next could be a path for others to follow.

Or a warning of what to avoid.

Much of this destruction is occurring in biological "hot spots," primarily in the tropics, where diversity is being destroyed at the greatest rate. Vietnam is one such naturally occurring hot spot, where the tropics meet the temperate zones, mountains meet the sea, and isolated pockets of wildlife were marooned during the last ice age.

Protecting animals in a few key places such as Vietnam could be instrumental to preserving the world's genetic diversity. Scientists in the peer-reviewed journal *Nature* said that by saving the Vietnam hot spot and twenty-four carefully selected others, we could save the greatest variety of wildlife while spending the least money.

The world environment is changing so fast that there is a small window of opportunity that will close in as little time as the next two or three decades.

The EPRC staff tries to make things as natural, homelike, and familiar as possible to their animals. Some cages/enclosures are thirty feet long by fifteen feet wide by about twelve feet high; all have screened ceilings, concrete floors, and lush native vegetation inside and out. A pair of Australian volunteers, Susan Rhind and Murray Ellis, explains to me that a new trend in the husbandry of wild animals has been fostering "cage enrichment." Much like the approach of the San Diego Zoo, this means that even small spaces incorporate springy branches, roots, rocks, stems, and overhanging vines that extend in every direction to give a three-dimensional quality to the interior, along with lots of vegetation. The era of the square, empty box with a bare floor and iron bars is disappearing. In addition, for some species, rather than dump the food directly into a bowl, zookeepers hide it in various places around the cage/enclosure, making the animals hunt and forage for food much as they would do in the wild. The activity helps to keep the creatures alert and vibrant.

These enriched environments can seem remarkably like the real thing. To give me a sense, one early evening, Susan Rhind shows me into a cage/enclosure at the Civet Rescue Center next door—one of the many spin-offs from Nadler's work at the EPRC (along with other projects)—and shuts the door firmly behind me. Though I was completely within the fenced confine, it felt like being in the depths of the jungle, particularly when I discovered that its occupant, *Chrotogale owstoni,* a type of wildcat that hunts at night, was still inside. The civet was harmless, and more interested in the lizards and bugs that lay under a log than in a human intruder. Still, I was glad when Rhind opened the cage door and let me out.

Every feature inside serves a purpose: besides cage enrichment, they have concrete subfloors that make for easy scraping-up of dung, sweeping, and hosing down, while also making it hard for anything to dig its way in or out. Cyclone fencing on the ceiling ensures that the animals can't escape from above, nor predators and poachers sneak in. Between each cage/enclosure, raised brick walkways connect all the various buildings, sheds, and other structures, allowing for easy access in the depths of monsoon season. Every enclosure has a double set of doors to prevent escapes. In the event that something does get out, the entire area is ringed with a double set of fencing and double sets of gates. Security is the watchword.

However, if there is one element to the center's success with langurs that is key, it would have to be the food supply. As sensitive and fastidious as a Siamese cat, langurs eat leaves from these limestone mountains that are high in cellulose and sometimes even toxic; this requires them to have a complex, multipart, specially adapted stomach containing bacteria that make it possible to digest leaves, seeds, and the occasional unripe fruit.

Their unusual stomach arrangement is something akin to that of a cow: the four compartments of a cow's stomach break down tough, stemmy materials in a process that involves digesting grasses, fermenting them, partially regurgitating the resulting mass (the cow's "cud"), and redigesting it. This rough analogy may be a gross simplification of what happens inside a langur, but the basic idea behind it is similar: a specialized, compartmentalized digestive system allows for eating the otherwise inedible, and so its owner must be very particular about what it eats. In fact, if a langur has been fed too much ripe fruit, the bacteria and enzymes in its stomach cannot handle it. The langur succumbs to vomiting, diarrhea, and weight loss, deteriorating rapidly, until the creature dies.

Consequently, every day at the EPRC, seven trained "cutters" go into the jungle to harvest great wheelbarrow-loads of leaves for the langurs: three different kinds of vegetation native to North Vietnam, of 220 pounds apiece, or about three times what had been needed the last time I visited. A regular early-morning ritual each day is for a team of a dozen people to sit down and then sort out the leaves and distribute them into equal portions, like a Vietnamese version of a corn-husking bee. It costs $1,000 in labor costs per month just to feed the langurs alone, and this in a country where the dollar is extremely strong and wages are low. (Animal keepers make between $100 and $150 per month; many Western tourists can get by in Vietnam on $10 per day for meals.)

When you walk into a cage with a finished bundle of leaves destined for one of the langurs' twice-daily feedings, it can be a bit of a rodeo. You have to be alert and on your toes, especially as some larger enclosures contain groups of several of the creatures. One Cat Ba langur in particular likes to steal the padlock from the inner door and try to get out; he's made it as far as the entryway between the inner and outer doors a few times. Says Jeremy Phan, an American primatology student doing an internship here as part of his master's degree at Ohio University: "Langurs are really smart. It's funny, but it can be a pain."

And while you're keeping one eye out for the Cat Ba's latest tricks, the other langurs have already spotted the youngest, freshest, tastiest, and most desirable of the leaves you are carrying, and they follow you, trying to steal branches out from under your arms while constantly jockeying for position for the upcoming meal.

Once they've stuffed themselves full, the langurs must sit for a long time while the all-important digesting takes place. During this time, the animals grow lethargic, and they look like nothing so much as a bunch of contented cats full of cream, sprawled over the furniture. The langurs' stomachs become huge and distended pot-bellies while they work on processing this very fibrous diet.

Each animal looks like a snake that has swallowed a beach ball.

Gazing at the scene, Phan commented, "A whole lotta leaves go through this place."

When we get back to the office, Tilo and I talk some more about the progress of the EPRC and its goals. He's for never throwing in the towel, arguing that no one really knows at what point a species was "functionally extinct"—to use the dry, scientific terminology. Was it when a species was down to its last dozen individuals? Or the last hundred, or the last thousand? A lot of creatures, I was to learn, have made a successful comeback from very tiny populations, escaping the consequences of squeezing through that genetic bottleneck. Scientists are still answering what "functionally extinct" is.

As the afternoon wears on, the conversation turns to the intensely personal reasons behind why individuals decide to study and save species—over and above the more rational, dry logic. Some biologists, such as E. O. Wilson, use the term "biophilia" to explain the deep, sometimes subconscious connections that humans often feel toward other forms of life, expressed as the urge to be outdoors and watch animals—with protecting the animals as a natural consequence.

This idea was echoed by Bill McKibben, a well-known environmental activist, "deep ecologist," and scholar-in-residence at Middlebury College, Vermont. "If you step back far enough, *nothing* matters," he replied to my query. "The Sun is going to blow up in a billion years and then we're all extinct. But in the meantime, it seems right and proper to try and succor the flora and fauna we

were born onto the planet with. We're all creatures of the late Pleistocene, and that should be enough. At least, it is for me."

And this "biophilia" is not just limited to animals. A botanist I
know at the National Tropical Botanical Garden in Hawaii, Steve
Perlman, specializes in rescuing the world's rarest and most endangered plants, and in the process rappelling down cliffs that a mountain goat would find challenging. Sometimes he and his colleagues
leave a plant in the wild, when it seems too risky to be moved.
After years and years of seeing the same plant in the same place, it
becomes a part of his life—and painful when the plant dies. Perlman said, "I've gone back and actually witnessed extinction at least
a dozen times. And then I think, yeah, I'm not coming here again.
I'll go out and get drunk or something, because I've just lost a
friend."

When I had pitched the question to George Schaller, he considered saving fauna and flora to be an antidote to sitting around and
fiddling while Rome burns—an attitude that I heard repeatedly
from researchers in Vietnam, regardless of their nationality. Pas-
sively watching something become extinct was too depressing, and
he preferred to at least try to do everything he could think of to save
a threatened or endangered species—even if the odds sometimes
looked hopeless.

"In the past fifty years, more natural resources have been used
up than in all previous human history," Schaller stated. "But this
doesn't mean that I personally go around moaning. I take something very specific, where I feel I can make a difference, whatever country it's in. And I've been working in particular in China,
where they're very pragmatic, and if I come up with good ideas,
they actually try to implement them."

Schaller said that of all the wildlife rescue projects he's done in
his six-decade-long conservation career, he thought his work on the
Tibetan plateau had the greatest influence: with a lot of encouragement and prodding, the Chinese government set up about 130,000

square miles of reserves there. When he first started talking about conservation there, local officials looked at him blankly. Now they point to genuine, real rescue programs put into place.

In particular, Schaller cited their combined efforts to save the Tibetan antelope—a big undertaking. "Tibetan antelope migrate long distances, so to protect them you have to protect the whole landscape. And they've got the finest wool in the world, in great demand. During the decade of the 1990s, a total of about 300,000 were slaughtered and the wool smuggled to Kashmir, where the weavers made these very expensive 'shahtoosh' shawls for the wealthiest women in the world—so fine that an entire shawl can fit through a wedding ring. With a publicity campaign, I raised awareness of how the animals were killed to make the material for the shawls, so there's much better protection now. The trade still goes on, as it always does—but it's way down and the antelope numbers tripled in ten years."

Schaller continued: "It's funny but I started out just to be outdoors and watch animals—it's a good way in the sciences to make a living. But then you realize that you have a moral obligation to help what you study. You get some perspective as you get older: you've got to do something in this life, to help society in some way. And conservation is about the most benign thing you can do to help your country and the world."

But probably the most poignant personal reason for rescuing endangered wildlife came from Alan Rabinowitz. When asked what drew him to his work, he replied that, at first glance, wildlife biology would not seem to be an obvious choice for a born-and-bred New Yorker from the Bronx. But Rabinowitz had been a bad stutterer as a child, and early on he had discovered that talking to animals allowed him to temporarily overcome his impediment.

"Being with them put me in a relaxed situation, where I could work through the difficulties of speech," he said. "So, from kindergarten to sixth grade, I would come home from school and take my hamster and my turtle into the closet, where I could talk to them."

As he grew older, Rabinowitz progressed to bigger and bigger creatures. "Every weekend my father would take me to the Bronx Zoo, which had the 'Great Cat House' in those days—with the old-style cages with the iron bars and the bare concrete. And I would watch this one lone jaguar that I felt that I could really relate to—it had all that power and grace locked inside, which was how I felt. I loved going there, and talking to it. And over time, I realized that animals can think—they have feelings and consciousness and awareness, but we're not aware of it because they don't have a human voice. So, in my childhood, as they were allowing me to pour my heart out to them, I made a promise: if I could ever find my voice and control my stuttering, I would try to be their voice. I would try to speak for them, and save them."

Rabinowitz went on to work for the Wildlife Conservation Society, which is closely allied with the Bronx Zoo, and he created the world's first jaguar reserve, located in Central America. He overcame his stutter to become a prominent spokesman for the WCS.

As for what drove me to report about these conservationists' endeavors, I traced my curiosity to childhood evenings spent watching reruns of the old *Wild Kingdom* show on the family black-and-white television set. "Reality TV" before there was such a term, the program chronicled the efforts of a pair of zoologists, Marlin Perkins and his sidekick, Jim Fowler, to collect new specimens for the Saint Louis Zoo. Marlin was the silver-haired, mustachioed, impeccably dressed older man who always got the most camera time, but I always found myself secretly cheering for Jim, the poor slob who actually had to do the dirty work. Jim seldom had a line on-camera but always seemed to be the one doing things like leaping out of a helicopter onto the back of a wild elk, while Marlin did all the talking—usually from a safe distance away.

In one memorable episode, the duo went off to collect a twenty-five-foot-long river-dwelling snake called an anaconda (*Eunectes murinus*) in the wild. While Marlin cheerily prattled away onshore that this species was "the biggest snake in the world, more feared by the local Indians than any other South American giant, fully capable of killing and swallowing animals as large as an adult human being," Jim was actually in the river wrestling with the beast in an effort to put it into a cage. As the camera rolled, you could see his increasingly desperate struggles. At one point the snake coiled itself around Jim's neck and began to squeeze, all while the incongruously upbeat voiceover continued. As Jim's face turned red and his eyes bulged, you heard: "The anaconda exerted fantastic pressure . . ."

It all made a tremendous impression on someone in primary school. From such things do lifelong interests spring. (Jim survived.)

Apparently this same TV program had a similar effect on others; in his memoirs, one researcher said that watching the giant-snake episode on *Wild Kingdom* as an eight-year-old was critical to his life's ambition to become a wildlife veterinarian. "I didn't know why he needed to catch the anaconda, but I knew I wanted Jim's job," wrote William B. Karesh, who now directs the International Field Veterinary Program for the Wildlife Conservation Society.

For me, this fascination with natural history and wildlife biology culminated in a much more modest way, in the form of two summers working in Yellowstone National Park, Wyoming—the world's very first national park, started in 1872—where I spent some of my free time as a part-time volunteer ranger/naturalist before succumbing to the lure of journalism.

Consequently, when reports began to emerge of Vietnam's newfound wildlife, the potential story exerted a tremendous pull: there was the appeal of covering the discovery of large, unknown, extraordinary creatures in an exotic land. Any newspaper articles about them were saved in a folder of dream assignments; I found myself caught up in the animals, the people, and the place. Even after the stories were filed, I couldn't help but continue to follow the animals and the epic struggle to save them—even if it was, too often, from far away.

As for Tilo, he absolutely does not like talking about any of these things, let alone delving into what drove him in his middle age to pull up his roots and move halfway around the world to a foreign country to save wild animals.

Instead, he prefers for me to see and experience the creatures and their natural environment for myself. Now that I've toured the EPRC and seen the langurs in captivity, Tilo makes arrangements for me to go with him to one of the places where they've been releasing some healthy langurs into the wild, at a wilderness reserve called Van Long. It's a bit of an undertaking, and won't happen for a few days. While he's coordinating the details, he and his wife arrange for me to take a little walk in the jungle of Cuc Phuong.

6

Vietnam's "Lost World"

This is just the tip of the iceberg . . .
there's an entire lost world out there.

—Oxford University researcher John MacKinnon

WHEN YOU'RE WALKING on an unmarked jungle trail in Vietnam, no matter how short the hike, the last thing you want to hear from your guide is "I can't read a map."

Our leader, Jeremy Phan, an American primatology grad student, was taking me to see one of Cuc Phuong National Park's famous giant trees; with us was Isidore Rionduto, a specialist in recording monkey vocalizations.

We had arrived at the site of the tree, about a twenty-five-minute walk from the paved road in a sizable chunk of old-growth forest. Along the way, we stop to get a feeling for the size of the trees; one is so big that when I wrap my arms around the trunk, my fingertips do not meet. After reaching our goal, the thousand-year-old "Ancient Tree"—most likely a specimen of *Terminalia myriocarpa,* or East Indian almond, although the printed description on the map was hard to decipher—and trying to capture in a photograph its fifteen-foot circumference, we had decided to take what looked like a shortcut back. It turned out to be longer than we thought, and the increasingly overgrown shortcut gave the appearance of being a mere collection of animal tracks.

Then Jeremy made his revelation.

Luckily, one of us had a compass buried in the contents of his backpack and we were soon able to confirm that the unmarked pathway, although winding all over the place, was generally heading more or less in the direction of the spot where we had left our bicycles next to the asphalt road.

The experience forces us to pay more attention to our surroundings.

Even though we're still a bit concerned about getting lost, it is impossible not to be impressed—which explains why Nadler was so eager for me to experience this immediately adjacent forest in person, as opposed to sitting in an office and hearing him describe the context in which rescue work is done. Such jungle is the setting for the biological gold rush in Vietnam, telling much about the conditions in which researchers work. I can only wonder what it must be like at Van Long, a twenty-minute drive outside the park, where the EPRC now releases those langurs that stand a chance at surviving in the wild. (Upon hearing the news about our upcoming trip out to Van Long, Hien got a reverential look on her face, sighing about how this much wilder place was her favorite. From what she and Tilo have told me, it sounds like Van Long is in line to formally become a protected wilderness area or national park all by itself, and hopefully without the problems that Tilo described earlier.)

Even in comparatively busy Cuc Phuong, one can immediately see how it would be easy for a creature to escape detection. In addition to several giant trees, there are five layers of canopy in this patch of jungle: there are plants underfoot and plants up above; there are plants dangling in the spaces in between. There are plants strangling plants, and parasitic plants feeding on other plants. There are woody liana vines of the type that Tarzan used to swing on. There are beautiful orchids—"epiphytic" plants that grow harmlessly upon the branches or in the crotches of the trunks of other plants, taking nothing from their hosts but instead getting all the moisture and nutrients they need from the air.

The forest has the greenhouse-like, moist smell of wet leaves and an overall primeval feeling. Part of this is due to the occasional presence of a cycad, a member of a plant family that has been around since the Mesozoic era 230 million years ago, when they shared the planet with the dinosaurs—the plants flourished so well during that period that some botanists call the Mesozoic the Age of Cycads and Dinosaurs. There are twenty-four species of these plants in Vietnam, a record number that makes this country the most cycad-rich in Asia. They stand out for their distinctive swollen bases, crowns of palmlike leaves, massive cones, and occasional seed-bearing leaves.

But while looking up at the treetops, we find that we also have to pay attention to what lies beneath our feet. We discover we have come to an area honeycombed with sinkholes—the limestone foundation of this countryside is readily dissolved by the mild acid created whenever rainwater mixes with decayed leaves or other rotting vegetation. The resulting chemical reaction leaves behind enough hollow pockets that portions of the underlying geology resemble a giant Swiss Emmentaler cheese. One sinkhole, just a few feet in diameter at the surface, apparently leads to a larger cave complex; I can feel a slight cool breeze coming up out of the dark side passageway below when I stand on the sinkhole's lip. It turns out that such breezes are a classic indicator of a cave's presence. The breezes are the cave's "breathing," caused by warm and cold air being exchanged with the outside world as the earth alternately heats up in the day and cools down at night.

Due to all these holes, any liquid is quickly funneled into natural underground drainage systems—traveling so fast that there is very little water left above to form ponds, lakes, or running streams. Often, the surface water simply disappears partway down the steeper hillsides, emerging in the form of bubbling springs on the lower slopes.

Consequently, Vietnam's limestone mountains and limestone-based forests are pockmarked with water-eroded caves. About 265 miles from here, in Phong Nha-Ke Bang National Park, there are more than three hundred caves, including the world's largest, Son Doong Cave. The latter is so big that an entire block of forty-story buildings could fit into just one of its many chambers; a complete 747 airplane could be parked inside another chamber with plenty of room to spare.

In 2009, a British-Vietnamese caving expedition found that one passageway of Son Doong was so tall that clouds were forming near the cave ceiling. Ninety miles of Son Doong have been mapped so far, with much of the cave system yet to be explored. All kinds of life have been found within it, including fish, wood lice, and millipedes; all are a ghostly white, a common feature when many generations have been born, bred, and died in a world without light.

Humans have also lived in such caves; during the Vietnam/American War, thousands of North Vietnamese hid in caves and underground bunkers during B-52 bombing raids by the US Air Force. I later visited one such underground hideout near the old demilitarized zone that had been the dividing line between North Vietnam and South Vietnam; it was so elaborate that it even contained an underground hospital, where seventeen babies had been born. In the early 1940s, during World War II, Ho Chi Minh had planned his strategy against the Japanese and the French from Pac Bo Cave, north of Hanoi.

Watching the ground below and the vegetation above, we eventually find ourselves on the paved road a few hundred yards from our starting point. After this experience, we stick to the marked trails for our next destination, Dong Nguoi Xua, or Cave of Pre-historic Man.

We had come to this part of the park because Nadler had wanted me to experience for myself a slice of Vietnam's "Lost World," and learn what kind of a place existed here before rapid development had come along. I knew that such things as "biodiversity hot spots" existed—places especially rich in endemic species and most threatened by human activities—and that Indo-Burma was considered one of these hot spots. More and more, however, I found that there was not one lost world, but several: bits and pieces scattered here and there that have long been rich in variations of life.

In some cases, there was hot spot within hot spot, decreasing in size as you got more rigorous about your parameters. Defining the thing I had come to see was something like pulling apart a Russian nesting doll: first, there was the generalized hot spot of the tropics; then more specifically Indo-Burma; then Indochina; then Vietnam; then northern and central Vietnam; then the caves, islands, limestone karsts, and other unique landforms of this country that act as nurseries of biological diversity. (Of course, it's possible to carry this reductionism too far, to the point where something is cut so fine that it has no visible function left. I remember dissecting an animal brain once in high school biology class, in an effort to get at just what a brain is. I kept dividing it into smaller and smaller pieces, until there was nothing left but a formless pile of sliced-up gray matter on a tray.)

In the case of Vietnam, it seemed that the greatest biodiversity these days was found on the margins—forgotten bits, not well suited to agriculture, usually located in more inaccessible, difficult, or otherwise undesirable country. It says something about the dogged persistence of nature that life can survive and sometimes even thrive under these conditions. The phenomenon is sort of an Asian version of the ailanthus trees (*Ailanthus altissima*) stubbornly growing in the vacant lots and between the pavement cracks of Brooklyn, where their trunks grow and resprout despite being cut down or even set on fire. The ailanthus was the species at the heart of *A Tree Grows in Brooklyn;* author Betty Smith saw its tenacious survival as a metaphor for life on the streets, describing it as "the only tree which grew out of cement." Rounding things out, the ailanthus, alias "the tree of heaven," is a native of China, first brought to the United States as an exotic ornamental in the 1700s, before it became an invasive weed species.

But while many species can thrive in neglected, overlooked, out-of-the-way spaces, they do not do so well in spots where the soil is constantly, repeatedly plowed over without respite. Because the southern part of Vietnam was so intensively farmed for so long, it does not have that many wild species left. The land in the area dominated by Saigon was the most dramatically transformed to grow rice. The Mekong Delta's natural areas were fragmented by ever-expanding spiderwebs of canals dredged to drain the region and provide transportation between central administrative areas and outlying villages. An estimated 1,500 miles of canals have been constructed over the centuries, altering natural flood patterns. Despite this, some areas have staged a comeback, with the restoration of a part of the Plain of Reeds and the return of its Eastern Sarus cranes as a prime example.

But overall, thousands of acres of wetland became rice paddies, devoid of much else. During the French colonial era, the amount of land under rice cultivation in the Mekong quadrupled between 1880 and 1930. To wildlife biologists, this means that a lot of southern Vietnam is now "just one big rice paddy," as Martha Hurley described it, where little native wildlife remains. As a result, it attracts fewer wildlife biologists than northern Vietnam—but some still make the effort there.

To get a better understanding of why northern and central Vietnam are so different from the southern portion, it helps to take a good look at a map. Vietnam's overall shape loosely resembles a set of barbells placed on end: there is a large, bulbous expanse of land in the north centered on the Red River Delta and the city of Hanoi; then an extremely long, thin, narrow stretch of land running north-to-south that blends seamlessly into the high plateau known as the Central Highlands; then another, not-quite-so-bulbous expanse in the south radiating out from Saigon—now Ho Chi Minh City—that includes most of the Mekong River Delta.

Human settlement echoes this barbell shape; the big river deltas at each end are well suited to rice cultivation and so are home to the largest human populations. (Some observers have said that Vietnam's geographic shape resembles a giant carrying pole with a basket of rice at each end.) The area between the deltas holds fewer people, their number diminishing as you go from the lightly populated Central Highlands near the central coast to the near-empty parts of the Annamese Cordillera in the interior, also known as the Annamite Mountains.

Consequently, the narrow, central-most portion of the Annamites, where steep mountains nearly reach the sea, has always been a convenient place for geographers, statesmen, and others to separate the north of Vietnam from the south. At its narrowest point, near Hue, Vietnam is barely 30 miles wide. At its widest, Vietnam is more than ten times that size, or close to 380 miles wide.

More than four centuries ago, the portion near Hue marked the boundary between the rival ruling dynasties of the Trinh in the north and the Nguyen in the south, and was where the two rival clans fought for control of the whole country. Some historians saw parallels to these ancient North/South, Trinh/Nguyen power struggles in aspects of America's war in Vietnam in the 1960s—and perhaps a clue to the lingering animosities within the country today.

When the French colonizers arrived in force in the

mid–nineteenth century, they took advantage of the country's geographic and cultural distinctions. Under a strategy of divide and conquer, they ruled the north as a separate district that they called Tonkin, and the south as a district they called Cochin. French geographers declared the area between the two regions an entirely separate entity that they called Annam, which they found useful to demarcate as an entity that ran for hundreds of miles up and down the coast.

No matter how broadly its boundaries were defined, the central portion of Vietnam was always steeper, narrower, rockier, and less populated than the major power centers. It's also much wetter in this region than elsewhere; Hue is infamous for rain, as I learned to my dismay. American field biologist Shanthini Dawson tells of experiencing ten days of rain in a row without a five-minute break west of Hue in a nature reserve in central Vietnam, near the Laotian border.

Vietnam's most mountainous areas lie in the northern and central regions; the mountains of the Annamese Cordillera continue along the border with China and eventually run all the way to the Himalayas. Many of the country's northern plants and animals are closely related, evolutionarily, to those of the Himalayas and southern China. For example, some of the closest relations to the turtles of the Hanoi region are found in Iraq, Syria, and Iran; the "pattern is likely due to separation of their common ancestral population by the rising Himalayas," says one biological reference work.

Historically, neither northern Vietnam nor its central Highlands were that well known to science, especially when it came to wildlife. There was a big collaborative, joint scientific expedition to these regions in 1930, called the Kelley-Roosevelt Expedition, which incorporated the work of a noted ornithologist in Indochina named Jean Delacour. (Birds are generally easier to observe and identify than mammals: they are usually out during the daytime, have species-specific songs that help identify them, and are often brightly colored—making them easier to spot. Consequently, many surveys in Indochina concentrated on the region's birdlife instead of its mammals.)

To modern ears, the members of these expeditions sound like a bloodthirsty lot. Consider this field report from the Kelley-Roosevelt and Delacour Asiatic Expeditions of 1930, in which the participants describe collecting Ha Tinh langurs in Vietnam for Chicago's Field Museum of Natural History: "A second later I saw another and brought it down with two shots. Then Kermit started shooting just beyond me, and for a few minutes it sounded like a miniature battle as we fired at half-seen shapes flitting through the tree-tops." (The Kermit referred to is Kermit Roosevelt, son of US president Theodore Roosevelt.)

Other than these field collectors, the early researchers into Vietnam's fauna could be counted on the fingers of one hand. The Annamite Range, particularly above the country's narrow "waist," was considered just too wet and too remote to explore using the methods of the time. Of the few scientists who went out into the bush, nearly all were non-Vietnamese; this was during the height of the French colonial empire in Indochina, when the rubber plantations, coal mines, rice paddies, tin mines, and the official, colonial government–run opium monopoly were all operating in full swing. (An enterprise similar to that of the British East India Company's opium-growing concern in India, the French colony's opium business accounted for between 15 and 33 percent of the colony's tax revenue from one year to the next.)

It's a pity that so few scientific surveys or wildlife census counts were done, because recent work shows that Vietnam is a country with high "species-richness." It is now considered among the top twenty-five countries in the world in terms of the number of mammal, bird, and plant species per square mile.

Similarly, Vietnam's flora is rich for the country's size. More than ten thousand vascular plants are known, and scientists estimate at least another three thousand exist. But however many species there are in the country now, it is likely that neither central Vietnam's forests nor those of neighboring Cambodia can hold a candle today to the diversity or the numbers of the early twentieth century, when big-game hunters described the wildlife abundance here as second only to the Serengeti's. On individual hunting trips, they recorded respectable numbers of what are now extinct large mammals such as the wild water buffalo (*Bubalus arnee*). These hunters apparently also saw much larger numbers of Asian elephant, Eld's deer, Javan rhino, Asian black bear, gaur, and tiger than visitors can ever hope to see today.

By comparison, in our era, the number of Asian elephants in Vietnam is measured in the tens instead of the hundreds, the IUCN does not even know if any Eld's deer still exist in Vietnam, the Javan rhino may be extinct on the Asian mainland, Asian black bears are taken from the wild for bile farms at a rate that is quickly decimating the wild population, and the gaur declined by 90 percent in Cambodia in thirty years. (The last tally of gaur in Vietnam was fifteen years ago, and only a handful were found in the country at the time.) As for tigers, ten years ago there may have been sixty or seventy of them in Vietnam, said researcher Nguyen Manh Ha of the Center for Natural Resources and Environmental Studies at Vietnam National University in Hanoi; now he estimates the number is closer to twenty or thirty.

Because there were no regional, comprehensive wildlife surveys of Vietnam, and few wildlife studies of any kind, it is hard to pin down exact figures on the original number of game animals countrywide, or the exact percent of decline. Concern about wildlife was in its infancy in the region before the wars of the twentieth century and the current postwar economic boom. It is hard to report a decline if no surveys, or minimal surveys, were done in the first place.

But hints can be gleaned from the occasional statistics reported in specific regions, and those examples can give an idea as to what the wildlife situation may have been nationwide. As late as 1920, in a guide to hunting in the resort/safari area of Dalat, in the cool uplands high above Cam Ranh Bay about two hundred miles northeast of Saigon, the regional head forester boasted: "The Lang-Bian's hunting grounds can be compared to some of East Africa's in their abundance and diversity of animal species . . . one can be assured of running into some large game."

Dalat and the plateau on which it sits—Lang Bian—are as high as seven thousand feet, with a cool climate and a vegetation that gives the area an Alps-like feel, where even today, tourists revel in the need to use a blanket at night after the sweltering torpor of the coastal lowlands. Sometimes called "Little Switzerland," the resort was deliberately created from scratch in raw wilderness in the early 1900s as a rest and recuperation area for colonial officials. It was filled with chalets that gave the spa-like feel of a fashionable Beaux-Arts, continental Europe in the midst of Asia, much like what the British did with their "hill stations" in India at the height of the Raj. (Creating such an environment came at great human cost. A 1908 book, *Champoudry chez les Moïs,* said that about twenty thousand "coolies" died in the construction of the road that connected the mountains of Dalat with the seacoast.)

Dalat was the unofficial capital of Vietnam in the dog days of summer then, where government administrators and their families went to escape the heat and indulge the fashion for upscale hunting safaris. To give it an atmosphere more like "home," French pastry stores, butcher shops, and restaurants were started. To supply them, Western plant species were introduced; some transplants did quite well, which is why a person can eat fresh strawberries from Dalat in Ho Chi Minh City to this day. As early as 1903, the region was growing such nonnative, Western staples as onions, potatoes, carrots, celery, beets, parsley, tomatoes, turnips, and green beans.

No one has apparently charted the effect of these foreign imports on the environment of Vietnam today, but American grass, Australian eucalypts, and other introduced exotic species continue to plague environmental restoration projects. But the appeal of a little Europe in the midst of Asia continues, and an American tycoon restored the town's most sumptuous Beaux-Arts colonial hotel in the 1990s to provide a sanitized, upmarket version of the past and capitalize on nostalgia for what the French call "Indo-chic."

In its heyday, hunting was the star attraction of the Dalat region, and "hunting lodges"—comparable to the "Cottages" of Newport or the "Great Camps" of the Adirondacks—were built to cater to the trade. One official gave a mind-numbing list of all the large mammals—tigers, panthers, deer, bears, gaur, elephants, and others—that sportsmen could easily bag as trophies. In his 2011 book describing the realities of that heavily romanticized era, *Imperial Heights: Dalat and the Making and Undoing of French Indochina,* Eric Thomas Jennings drew upon the contents of newly opened French government archives concerning the hill station; among other things, the previously secret files tell of a single hunter named Fernand Millet who alone killed 600 deer, 150 gaur, 50 tigers and panthers, and 40 elephants. Ironically, Millet doubled as head forestry guard for Dalat.

Look at one typical guidebook/photo album from Millet's time and you see a white lady in hunting attire and rifle, triumphantly standing next to a dead tiger she has killed, while peasants in traditional conical hats look on. Similar images abound, of hunters sitting atop dead elephants or otherwise posing with their slain trophies.

Even by the standards of the day, the extent of the slaughter was alarming, and there were reports of "coolies" refusing to transport safaris any farther after just one tiger had been felled—which was laid down to primitive superstition. However, fellow aristocratic hunters also began to recoil (and become concerned that there would soon be nothing left to hunt). The archives revealed that the governor-general of the French colonies in Indochina, Albert Sarraut, complained of "barbarous and irrational hecatombs" in the jungle, and that he decreed that a "reserved" hunting zone be set aside, where only male gaur, male buffaloes, and male elephants could be killed. Formal hunting codes backed by law were enacted in 1928, and the idea of national parks based upon the American model—Yellowstone was the world's first national park, and the US National Park system set much of the tone for the rest of the world—was studied.

If the experience at Dalat was repeated at the ten other official hill stations in Vietnam, plus similar enclaves in Laos and Cambodia as well as unofficial private enclaves, it is obvious that there was a great deal of wildlife destroyed from hunting alone. Factor in war, development, large-scale commercial farming, fragmentation of habitat, and a human population that has increased eightfold since then, and it's apparent that the original wildlife populations must once have been much bigger than they are today.

Anecdotal evidence, oral histories, and other sources support this: twelve-hundred-year-old Cham temple sculptures depicted an abundance of gaur, for example, and prehistoric cave paintings show large numbers of kouprey—the same animal also appears in bas-relief on the monuments of Angkor Wat. In addition, the logbooks of early-nineteenth-century ships tell of hunters bagging vast numbers of langurs off Da Nang in less than an hour. More recently, reputable sources such as the former director of Cuc Phuong National Park—Dao Van Khuong—said that he personally saw tigers in the park in the 1970s. And Nadler said the forest of Cuc Phuong is much quieter now than when he first arrived in the early '90s, when there was much more hooting in the trees from the gibbons and other wildlife.

At the same time, it is hard to know how much is for real and how much is inflated hunters' tales; for every park director or accredited field biologist, there is a horde of casual visitors who can claim they saw something. And paintings and sculptures are not the same as scientific surveys.

Nevertheless, there is a pervading sense of a dramatic decline. The situation is somewhat like going fishing off Cape Cod or the Grand Banks, now that the cod population has officially become "commercially extinct" from overfishing. You see the photographs of giant fish from the past, read the occasional old reports of massive catches, and hear the stories—when John Cabot explored the Atlantic in 1497, his sailors simply dropped baskets over the side to haul in as much as they wanted to eat—but unless you are careful, it is easy to dismiss it all as mere fishermen's tales. The tendency is to think that whatever you see or do not see in the ocean now must be the way that it has always been.

Consequently, it is hard to mentally go back in time and visualize what the natural environment must have once been like. In 1873, Alexander Dumas wrote, "It has been calculated that if no accident prevented the hatching of the eggs and each egg reached maturity, it would take only three years to fill the sea so that you could walk across the Atlantic dry-shod on the backs of cod." Cod were known to be extremely fertile. A female codfish of "middling size," which by mid-nineteenth century standards referred to fish merely two feet long, could contain about eight or nine million eggs.

It seemed impossible that such a prolific animal could suffer a population collapse, wrote Mark Kurlansky in his book *Cod: A Biography of the Fish That Changed the World*. "But after 1,000 years of hunting the Atlantic cod, we know that it can be done," he said. Kurlansky also opined, "It is harder to kill off fish than mammals."

Which may be one reason it is so gratifying to find some undiscovered stragglers today.

We arrive at the Cave of Prehistoric Man via a wide, broad path, a concrete bridge, and a series of stone steps. It's obvious that this is a very popular spot; there is a paved lot at the trailhead for vehicles to park, along with a small cart selling cold drinks. I fear I am about to be disappointed.

The cave itself is about the size of a small house, with a few interesting stalactites, piles of small red twigs, bits of paper, a scattering of leaves and flower petals, shiny pieces of glass, and other debris scattered here and there on the floor. Granted, the cave is a cool, refreshing spot, with a commanding view, a few hundred feet above the trailhead and overlooking a small stream, whose gurgling we can hear way up here.

But looks can be deceiving; there is much of significance that is subtle, which I easily overlooked at first glance. What first appears to be litter left behind by picnickers turns out to be small shrines. The red twigs are spent incense sticks, the plant materials are flowers, and the bits of paper are replicas of paper money, left behind as offerings to appease the earth spirits, or *tho than*. The pieces of glass reveal themselves to be small mirrors to ward off harmful forces.

Many Vietnamese follow a mixture of Taoism, Buddhism, and animism—the belief that objects in the natural world possess spirits or genies. In northern and central Vietnam, a traveler can see a number of small shrines dedicated to the local *tho than,* who often reveals himself by letting a single tree grow in an otherwise open field. No farmer or villager would even consider cutting down such a tree; instead, on auspicious days—such as before the harvest—they place offerings at the base of its trunk. As practiced in Vietnam, there are huge numbers of various *tho than* and other spirits that need to be constantly appeased. *Pagodas, Gods and Spirits of Vietnam* explains that these spirits are considered to be "sly, hypersensitive, vengeful and capricious. Tirelessly they lie in wait, taking advantage of any opportunity to do harm. However honest a person might be, however hard-working and thrifty, he is always at risk of crossing an evil spirit." It sounds similar to the multiple gods and goddesses of the pagan ancient Greeks and Romans.

Hence the many shrines and small offerings to keep these troublesome beings happy.

Even the most urbanized Vietnamese seems to incorporate this mixture of belief systems easily into modern life, praying to the spirits before building a new house, for example. Magicians and sorcerers are still consulted, even in metropolitan Hanoi today. (Lest we Westerners laugh, in our own building traditions, workers often mark the completion of the framing of a structure by ceremonially attaching a small evergreen to the top of the highest roof beam. Known in New England as "topping out," the practice is probably a remnant of the West's collective Druid past. I participated in such a ceremony myself, while helping to frame a house.)

In Vietnam, caves and grottoes are considered to be a favorite dwelling of spirits, so they are a good place to locate an informal shrine because they form a direct link between the surface world and the earth spirit below. One cave near Da Nang, called Huyen Khong, has a natural opening in its roof and a carved figure of a Buddha in its depths; once a day, near noon, the sun's rays penetrate this opening to illuminate the statue in the cave's dark interior.

I start to realize that there is a lot more to the Cave of Prehistoric Man in Cuc Phuong National Park than first meets the eye.

Additionally, this particular cave was the site of one of the earliest discoveries of human habitation in Vietnam. Excavated in 1966, the cave revealed human graves, stone axes, pointed bone spears, oyster-shell knives, and grinding tools that date to about 7,500 years ago. In fact, this whole Red River Delta is a very ancient landscape, long inhabited by humans. (The delta itself contains silt with iron and aluminum oxides, giving the Red River both its color and its name.) Archaeologists have more recently found evidence that pushes the date for the first human presence here back even further than the date for the Cave of Prehistoric Man, and they now think that humans were probably living in caves and rock shelters in these

limestone hills between 9,000 and 18,000 years ago, with some disputed evidence hinting at a date of 30,000 years ago.

Although the origins of Vietnam's first humans are still un-
clear, anthropologists generally believe that some of them were
Austronesians—descendants of peoples from the islands of the Pacific and Southeast Asia. When I saw a traditional communal house
of the Sedang ethnic group of central Vietnam, near the province
of Kon Tum, I was amazed at how much the structure resembles
those of the South Pacific. The enormous thatched roof looks like it
was airlifted whole from Fiji. (Visitors can walk inside one of these
Sedang structures at the sprawling Museum of Ethnology in Hanoi,
with its extensive full-scale reproductions of village life among the
country's fifty-four distinct ethnic groups.)

Whoever those first people in Vietnam were, they soon began
to develop the arts; about 3000 BCE, the Dong Son culture, near
the province of Thanh Hoa, started to produce impressive bronze
drums and ironworks. For years, academics assumed that they imported their metalworking technology from China; now there are
at least a few scientists who think that things were the other way
around—they say that the Dong Son developed it on their own,
and it was appropriated by the Chinese. It's probably something the
Vietnamese and the Chinese will argue over for a long time.

In any case, the Dong Son were apparently a seafaring people
who erected massive stone religious monuments, much like those of
ancient Polynesia. They hunted wildlife with blowguns and, later,
crossbows—much the same way some of Vietnam's ethnic minorities in the hill country do today. Over time, they domesticated
livestock and began to raise rice, which allowed for a more stable
lifestyle that involved coordinated social effort to run irrigation canals, dikes, and so forth. Much of this early Vietnamese culture was
later absorbed by the expanding Chinese empire.

As for the animals these first humans hunted, scientists have a
few theories as to why Vietnam proved so fertile a place for so many

new species to evolve. It was a naturally occurring biological hot spot, where the subtropics meet the temperate zones, mountains meet the sea, and isolated pockets of wildlife were marooned during the last ice age.

One theory, set forth by the zoologist George Schaller, director of science for the Wildlife Conservation Society in New York City, suggests that the Annamese Cordillera was a redoubt of rain forest throughout the last ice age, when climate shifts caused the forests to expand and contract.

During wet interglacial periods, animal populations might readily have ranged across these high, rugged areas, such as the mountains of Vietnam and Sumatra. "The Annamites are highly unusual," he told me, "because at one time they were connected to Indonesia, before the water levels rose."

During lengthy cold and dry spells, however, when the rain forests retreated, those populations could have become fragmented into discrete pockets, prevented from interbreeding by mountain barriers. "So, you had all sorts of species that were similar to Indonesia that got stranded in the Annamites," Schaller said. Each pocket of rain-forest animals would then have evolved independently of the other pockets, marooned on a fragmented "island" of warm habitat, cut off from the rest of its kind on the "mainland." The idea that Vietnam is home to the remnants of some kind of "Lost World" is perhaps not far from the truth.

Some support for Schaller's theory comes from the striped rabbit discovered in the Annamite Mountains a few years ago; it is closely related to another species found in Sumatra, 1,200 miles away. The same is true of the rhinos discovered outside Saigon, whose closest relative was in Java. The list goes on and on. No wonder that in its handbook to Vietnam, the World Wildlife Fund describes the country as "the Galapagos of Southeast Asia."

This concept of biogeographic islands of habitat in Vietnam is key, for evolution is speeded up on islands. The reason behind this can be understood in a simple analogy. Suppose three people find themselves stranded on a desert island, completely cut off from home, and there is only enough corn growing on the island to feed one of them. That means that two of them would either have to leave or die.

But suppose that one of the three decided to be a fisherman, specializing in feeding himself purely from the sea, without need of any of the corn. Now, the same island can support two people, because each is drawing on a different resource.

The same process would be repeated if the third person were to specialize in becoming, say, a hunter.

In each case, the carrying capacity of the island has increased, solely because the original three immigrants decided to split up into different specialties.

The same process is true for finches that landed on distant islands in the Galapagos or the Hawaiian Islands or any other places far from land—those that survive to pass on their genes are pushed by their environment to develop specialties to take advantage of the different food sources and avoid exhausting any one food supply. Finches that have strong beaks go for tough-to-crack nuts, while those with long, slender beaks pick at the succulent nectar of cactus flowers. Over time, strong beaks mate with strong beaks, and slender beaks with slender beaks. The competitive advantage goes to those offspring that have more and more-pronounced and distinct versions of these features, to the point where they have diverged so much from the first finch immigrants on the island that they can no longer interbreed. They are now separate species.

This is just a simplified example; our understanding of this "island biogeography," as evolutionary biologist Ernst Mayr christened it, has become so sophisticated that researchers can now factor in such variables as the distance from the island to the source population on the mainland, the size of the island, the number of founding individuals on the island, and the rate at which new immigrants arrive in order to make predictions about the number of new species.

As a result, scientists have learned that small specks of land can have large numbers of species brand-new to us. And familiar species can display brand-new traits, or lose old familiar traits entirely.

The idea of gaining a trait may make sense. But losing one? Why do the descendants of a cave-dwelling fish, for example, lose their coloration over the generations?

I was able to see a variation of this trait-losing phenomenon in person once on a small, Manhattan-sized island called Lord Howe, located in the South Pacific about halfway between Australia and New Zealand. One of its bird species, the Lord Howe woodhen, *Tricholimnas [Gallirallus] sylvestris*—descended from a species called the Pacific rail—had nearly died out. Then a pioneering conservation program launched by Island residents rescued the birds at the very last moment, raising their numbers from seventeen in the year 1980 to about three hundred today.

The Pacific rail was once a common species scattered across the Pacific, and recent archaeological evidence by scientists such as David Steadman of the Florida Museum of Natural History revealed that Pacific rails were once intrepid explorers and good flyers. In the course of conducting digs in caves and old lava tubes to look for common patterns in the arrival and loss of species, Steadman discovered the remains of prehistoric rails on every one of the numerous tropical islands he has excavated. The evidence showed that contrary to previous scientific opinion, instead of being confined to a few islands, rails were widespread, covering vast distances to colonize deserted islands in the Pacific, Indian, and Atlantic Oceans. "He's changed everyone's perceptions of rails on Pacific islands," said David Blockstein, head of the Ornithological Council in Washington, DC.

In addition, Steadman and others have found that rails readily adapted to alien environments and morphed into new forms, much like Darwin's finches. Each rail colony would eventually become a brand-new species—such as the Lord Howe woodhen—uniquely suited to fit the conditions they found on individual islands. Steadman guesses that before human contact there were about a thousand different species descended from one pioneering rail species.

But no matter how abundant or far-flung, nearly all of those first rails' descendants had one thing in common: for reasons scientists only now are deciphering with the aid of archaeology, computer modeling, and population biology, most of them quickly lost the ability to fly.

A key factor seems to be that there were no people, and no other predators, until very recently. By chance, Lord Howe Island remained undiscovered by humans until 1788, when the "First Fleet" of convict ships passed by on their way from Britain to found the penal colony of Australia. There's no evidence that any human had ever occupied Lord Howe, or had even seen the island, prior to that day. Unlike much of the Pacific, no Melanesians, Micronesians, Polynesians, or Australian Aborigines had ever set foot on it. "Lord Howe is located in a funny part of the Pacific, down in the southwest corner," states Steadman, whose own research into prehistoric birds and extinct species has led him nearly everywhere else in the Pacific. "Lord Howe somehow slipped through the cracks; no one got there," Steadman says. "It's a backwater. But that's what makes it so special."

When they first settled on Lord Howe Island tens of thousands of years ago, rails sustained themselves with small territories, allowing large numbers to survive on a small landmass, and they produced many young per year. Their colonization was aided by the absence of predators during what Steadman described as an "Eden-like" era. Even now, the island's atmosphere recalls an earlier time. I was able to stand at the foot of the island's Mount Gower, call to the Providence petrels (*Pterodroma solandri*) nesting above, and watch the birds glide down to land at my feet. At another part of the island, when I set my camera down, a Lord Howe woodhen pecked at the front of the lens.

When I told one old islander, Ray Shick, about my exploits, however, he was not impressed, saying that sometimes he could get the Providence petrels to land on his outstretched arm.

Biologists call this phenomenon "island tameness," the result of spending eons of time untroubled by predators.

Similarly, the bird also began to change physically; over time, the rail began to adapt to Lord Howe and become a full-fledged woodhen. Its wings shrunk and its breast muscles got smaller. Nigel Collar, a Cambridge University professor who studies threatened birds and is the author of several ornithological reference works, told me that this was a clever strategy in the evolutionary arms race: "A pair of wings is like a full rucksack. If you really are just going to go round the block, it's not energetically sensible to carry your camping things with you."

Scientists are not sure how long it took for a long-distance flying bird to evolve into today's flightless version, but population biologist Barry Brook of Charles Darwin University in Australia recently made an educated guess. Based on his computer simulations of woodhen numbers on Lord Howe, Brook estimated that it would have taken the birds several thousand generations, or a bit more than 10,000 years, to become flightless—an eye blink when measured on the scale of evolutionary time.

There was a downside, however. "When man gets to the island that your ancestors reached so long ago, he brings rats and cats and dogs, things that the birds never needed to escape from until then," said Collar in an interview. "And of course it's too late to grow your wings back."

The carnage from those first encounters was awful. An early ship's log described how sailors collecting fresh bird meat on Lord

Howe Island would simply stand amid a flock of woodhens and knock down "as many as we pleas'd wt a short stick . . . they never make the least attempt to fly away and indeed wd only run a few yards from you and be as quiet and unconcern'd as if nothing had happen'd."

As Charles Darwin once philosophized about such displays of island tameness: "What havoc the introduction of any new beast of prey must cause in a country before the instincts of the indigenous inhabitants have become adapted to the stranger's craft or power."

It was a scene repeated throughout the Pacific. In *The Song of the Dodo,* David Quammen wrote that 90 percent of the bird species that disappeared in the last four hundred years lived on islands. Miniscule Lord Howe Island may have lost more bird species than Africa, Asia, and Europe combined.

Fortunately, a captive breeding program came to the rescue just in time for the Lord Howe woodhen. There were only three breeding pairs of the bird left when the captive breeding program began. Luckily, the woodhen, still unafraid of humans and comfortable in its familiar habitat, responded well to the inevitable tagging and handling that's involved with being captured, fed, and bred in a tightly confined space. By 1984, researchers were able to release eighty-six captive-bred woodhens back into the wild. According to a paper by Barry Brook in the journal *Biological Conservation,* "It [the four-year-long woodhen breeding program] has since been described as a classic case for a successful captive breeding and reintroduction program."

In fact, the same principles of island biogeography can happen in any situation where there is a fragmented population cut off from breeding with its kin in the larger world. The idea has been extended to isolated caves, to green fertile mountains surrounded by empty deserts, and to lakes surrounded by dry land.

One of the more unexpected twists may be that this island

biogeography phenomenon can be applied to environments as simple as a mountain of limestone cut off from other limestone mountains by rice paddies, lakes, or savannah. Known as karsts, these eroded limestone hills—of which the Red River has so many—contain little available water and poor, thin soils. They are areas of high plant endemism, particularly among the orchids. Endemic mollusks, reptiles, cave-dwelling fish, and the leaf-eating langurs that feed on this specialized vegetation are also concentrated on these formations. Limestone outcroppings isolated by intervening lowlands often harbor different "species assemblages" on each island of suitable habitat. The plant life that grows over limestone is remarkably different in structure and species composition from other forest formations.

In fact, these limestone formations harbor more biodiversity per acre than other places. Two adjacent hills may have different plant communities that are as different from each other as two finches from two separate islands in the Galapagos.

The reason? The best explanation comes from the authors of *Vietnam: A Natural History,* who write that these limestone environments offer "peculiar, varied and stressful physical conditions."

> Limestone hills may be covered with sediments deposited by wind and water, but more often the soil is derived from the underlying limestone itself. The latter soils are thin, alkaline and poor in nutrients, with the exception of calcium and magnesium levels high enough to prevent some plants from acquiring other minerals and nutrients. Rainwater drains swiftly from karst surfaces; creating harsh, dry conditions . . . Shade in the lower stories is variable and incomplete. These factors combine to create highly varied microhabitats where soils range from poor to rich, thin to thick, and dry to wet, and the amount of light runs from dark to quite bright. This

variety allows limestone areas to support a broad spectrum of plants adapted to different conditions.

With a newfound respect for caves and karsts, we bike the twelve miles downhill to the park entrance. On the way, we see large numbers of butterflies on the roadside; there are about 280 species of them here in the park, and the annual migration is about to begin.

When we get back to the EPRC headquarters, I learn in passing that they've found yet another new species, a type of gibbon that they are naming the northern buffed-cheeked gibbon, *Nomascus annamensis*.

Appropriately enough, one of the ways that researchers were able to identify it as a new species was by analyzing the frequency and tempo of the gibbon's calls. The other, messier way to confirm their finding was by collecting droppings from the animals and reading the latent genetic information they contained about the animals' intestinal cells.

Such methods were the best ways to nail down a new species. Because they live high up in the canopy, gibbon identification is especially difficult—small differences in coloration are nearly impossible to make out from below, while tranquilizing them for identification purposes is out of the question, because there is no way that these endangered animals would survive a fall from such a great height.

If you are lucky enough to see one of the newly discovered gibbons up close—as in a pen at the EPRC—you see that it has a prominent crest, giving it a vaguely "punk" hairdo.

Later, one member of the team, Christian Roos, said that their discovery caused "a minor sensation." Their work was published in the *Vietnam Journal of Primatology,* with Nadler and Roos as two of the coauthors. And not all that long ago, they had also formally identified the Cat Ba golden-headed langur as a new species, as well as the François langur.

With so many new species and so many fine variations, it was hard to keep up. It seemed to happen all the time all over the tropics—from animals at the EPRC to plants at the National Tropical Botanical Garden in Hawaii. It was also sometimes hard to know what all the excitement was about; after all, if the animals and plants had been known to local people for generations, then why was it such a big accomplishment when something was formally "discovered" by Western science?

The best answer lay halfway around the world, and more than two centuries back in time.

7

At the Hands of the Demon Called "Science"

Something happens to the man who discovers a new species of animal. At once, he forgets his family and friends, and all those who were near and dear to him; he forgets colleagues who supported his professional efforts; most cruelly, he forgets parents and children; in short, he abandons all who knew him prior to his insensate lust for fame at the hands of the demon called "Science."

—Michael Crichton, *Congo*

THE EXCITEMENT OF the Vietnam researchers in finding new species was very real, and communicated itself to everyone nearby. I was to find the same thing when talking to botanists on the Pacific islands, whenever they discovered a new plant or rediscovered one that was once thought extinct.

"It's exciting, it's a bit of a buzz, really. When you find a spot housing some of the last of the gray-shanked douc's langurs, that's a thrill," said Ben Rawson of Conservation International in Hanoi.

But, with apologies to Michael Crichton, aside from competition and fame . . . why did they go to such extremes to name a new specimen? After all, the animals and plants were already well known to the indigenous people. So, what was behind the drive to find and label? If not money or bragging rights, was there something else at the heart of it?

The answers lay about as far from the jungles of Vietnam as a person can get: Uppsala, Sweden, at the home of the man who started it all. Son of an eighteenth-century preacher, his personality could almost be a template for today's field researchers: he was endowed with a love of travel and adventure, a willingness to learn about new species from indigenous people in exotic places, an urge to dress in regional costumes, an ability to quickly grasp the local languages, the desire to be the first in print—and a sense of pleasure in the unconventional.

His works were the precursor to such twenty-first-century efforts as *The Encyclopedia of Life*—"biology's moon-shot"—which is attempting to formally name and describe all 3.6 million (or 100 million, depending on whom you talk to) species on Earth.

In Sweden, his name is Carl von Linné. We know him as Linnaeus, and his system of nomenclature as the Linnaean system, bedrock of modern biology.

As biologists like to say: "God created, Linnaeus arranged."

Nowadays, it's hard to appreciate what a revolution he set in motion. We're taught in high-school biology courses, by rote, that all life-forms can be divided into a descending hierarchy of Kingdom, Phylum, Class, Order, Family, Genus, and Species, as if it were self-evident. But the system was far from obvious, and not arbitrary—despite the endless number of schoolchildren who had to memorize it with the mnemonic device "Karen Put Catsup On Fred's Gray Slacks."

Biologist Stephen Jay Gould once explained it this way: "Taxonomy [the science of classification] is often undervalued as a glorified form of filing—with each species in its folder, like a stamp in its prescribed place in an album; but taxonomy is a fundamental and dynamic science, dedicated to exploring the causes of relationships and similarities among organisms. Classifications are theories about the basis of natural order, not dull catalogues compiled only to avoid chaos." It's a description that taxonomist Colin Groves of the ANU would have to agree with.

For a dead white male whose name conjures up a dry and fussy profession, the real Linnaeus was colorful and lively, with a flair for showmanship. He also had a tangible love of his subject matter, especially plants. (Linnaeus once wrote in his journal: "I do not know how the world could persist gracefully if even a single species were to vanish from it.") He avidly collected newfound specimens, which were starting to arrive in great numbers at the ports of Europe from Asia, Africa, and the Americas during the Age of Exploration.

At the same time, Linnaeus had a puckish sense of humor, and knew how to express his ideas so that they had the greatest popular interest. Quite simply, Linnaeus gave botany sex appeal. He wrote of the everyday act of pollination:

> Love comes even to the plants. Males and females, even hermaphrodites . . . publicly display by their sexual organs which are males, which females. The actual petals of a flower contribute nothing to generation, serving only as the bridal bed . . . When the bed has been made ready, then is the time for the groom to embrace his beloved bride and surrender himself to her.

Linnaeus even went so far as to classify every plant by its sexual organs, labeling one *"Clitoria."* No wonder Linnaeus's regularly scheduled, show-and-tell walking lectures in the Swedish countryside each weekend were always popular among the undergraduate students he taught at Uppsala University.

But it was about more than sex.

One biographer wrote: "In addition to the first explanation that springs to mind, that Linnaeus was a kind of Peeping Tom of natural history, one can also identify a religious motif, that of Nature, in obedience to its Creator's call, being fruitful and multiplying."

What's more, Linnaeus's provocative lectures served a purpose: enlisting a new generation of adventurer-scholars to travel the globe on a mission to collect and classify. His students were so devoted that they called themselves his "apostles."

The religious connotation was appropriate, because Linnaeus thought that a "natural" system of classification could be arrived at by looking at God's original handiwork; this would have the benefit of showing how closely related any two plants are to each other, as opposed to arbitrarily classifying plants by, say, their medicinal properties. He felt that his proposed system would only work if it contained every plant and animal in existence; hence the need to go forth and collect.

As for categorizing what was found, he thought the system should be based upon a few specific, practical features that were simple to recognize and compare, even for amateurs. For a given plant, Linnaeus simply counted the number of male organs (stamens) and female organs (pistils) to determine what categories it should fall in. (His method for comparing animals was more complex but otherwise similar in design.)

However, his "sexual system" of classification raised hackles. A contemporary British naturalist, William Goodenough, was appalled by Linnaeus's "disgusting names, his nomenclatural wantonness, vulgar lasciviousness, and the gross prurience of his mind."

For there were serious disagreements. A rival scientist named Johann Siegesbeck called Linnaeus's method of classification "loathsome harlotry," and denied that plants could even have sexual reproduction. Linnaeus responded by citing the scientific evidence—and naming the ugliest, smelliest weed he could find the "Siegesbeckia."

And Linnaeus made sure to plant a sample of *Siegesbeckia occidentalis* in his own personal garden, said Magnus Lidén, a curator at the university's botanic garden and a professor in systemic biology.

Though some of Linnaeus's life sounds extravagant, everything is upheld by the historic record. A Linnaeus fan himself—there's a Linnean Society in London, and similar clubs in New York and Australia, among other places—Lidén took me six miles outside town for an off-season, behind-the-scenes tour of the family homestead, called Linnaeus Hammarby, now a national historic site and museum.

As we drove by fields still covered with April snowdrifts, we sometimes passed tall, large, flat stones set on end, with intricate serpentine carvings on their faces and obscure lettering painted in red. Typically placed at a crossroads, these were Viking rune stones.

Just before the turn to the Linnaeus homestead, Lidén stopped at a small, ancient whitewashed stone church. Even from the car, we could hear music coming from its interior—the church organist was practicing. Lidén knocked on the door and asked permission for us to look at the medieval ceiling inside. The organist said *"Jah"* and continued to pound away while we tried to make out the images above our heads.

To the accompaniment of thundering organ music, bright and cheery murals appeared that showed sea monsters, ogres, and devils happily devouring sinners—some of whom had fallen out of square-sailed longships. The images' creator, an early-fifteenth-century painter known to us only as Albertus Pictor, may have meant to warn viewers about hell, but everyone in his pictures seemed to be having a good time. Even the people being devoured appeared to enjoy themselves. Besides this church, Pictor did several others in this region north of Stockholm, and he is perhaps best known for a mural that showed Death playing chess with a human, the inspiration for a scene in Ingmar Bergman's *The Seventh Seal*.

Everything suddenly felt very . . . Swedish.

This was the same church Linnaeus once worshipped in; as a preacher's son, Linnaeus felt compelled to do the short walk from Linnaeus Hammarby on the occasional Sunday. A meticulous scholar, however, Linnaeus would leave if he found the sermon poorly reasoned or badly written.

Hammarby is an old Swedish word for "rocky forest slope," definitely the right name for the field next to the homestead. It looks like some old Scandinavian god dumped a load of rocks over the hillside; trees and plants scrabble to find space to grow between boulders. Nevertheless, what soil that can be found is fertile.

The scene is largely unchanged since Linnaeus's death 230 years ago. Built in 1762, the large, two-story main house is a combination of grand and spartan: the walls inside were made of squared-off logs stripped of bark, with moss stuffed in to fill the gaps; on them hang original oil paintings in gilded frames. The floors are bare of carpets or even a coat of varnish; they are simply long, wide, unfinished planks. There are a few ceramic stoves, but on the whole it looks and feels quite spare and chilly. Lidén glanced around and offered, "We had horrible winters before electricity and insulation."

One striking adornment was the master bedroom's "wallpaper." From knee height to ceiling, the bedroom's log walls were covered with large folio sheets of elegant, hand-tinted botanical prints of flowers, stems, seeds, fruits, and roots, drawn in precise anatomical detail. No two were alike. Marching in symmetrical rows over the walls, their combined effect was like some kind of optical art. Arranged for easy viewing from the bed, these botanical drawings were the first thing Linnaeus saw in the morning and the last thing at night.

A nearby portrait of Linnaeus, commissioned for his wedding day, showed a man with an intent expression wearing a wig and holding a book under one arm, while dressed in an expensive-looking, brilliant red coat and luxuriant, frilly shirt. Like many period paintings, this one was meant to display its subject's power and social position in the details. A key item lay in Linnaeus's right hand—a specimen of twinflower, or *Linnaea borealis*—his favorite plant. Described as small, humble, and commonplace, its attributes were the same ones that Linnaeus liked to use to describe himself. (Although for someone so humble, he wrote no fewer than four autobiographies.)

Another portrait showed Linnaeus as a vigorous young man just returned from a 1,200-mile, five-month scientific expedition in far northern Lappland. He came in triumph, with a cornucopia of new specimens and raw material for what would become a bestselling travelogue. To celebrate, he commissioned this portrait, which shows him dressed in the traditional reindeer furs of the nomadic Lapp—now Sami—people, clutching flowers and drums. Linnaeus was showing off his trophies.

Linnaeus may have had a penchant for self-glorification, but he genuinely was one of the first naturalists to do firsthand field research, collecting information about the fauna and flora of a distant land directly from those he met. Such "ground-truthing" speaks of something that some modern biologists, such as George Schaller, fear is being lost. Basic natural history is unfortunately going out of style these days. "It's all GIS analysis on computer, DNA studies and so forth," lamented Schaller. "But the fact remains that you have to get out into nature to look around and observe—something you can't do by sitting in an office with a computer. Don't get me wrong, the technology is great, and very valuable. But you've got to be careful that you don't get cut off from your real focus, and become a tool of your tools."

In contrast, Linnaeus immersed himself in the places he was studying, with an insatiable greed to collect, name, see connections and relationships, and to ultimately understand what he called "the simple plan of nature." In this way, he was much like Nadler or

Polet or any number of researchers in Vietnam. Linnaeus could no more walk away from a half-completed mental scheme of the organization of nature than others could leave a crossword half finished.

The need for a workable naming system was made even more urgent by the huge number of plants and animals being discovered at the time. Born in an era of constant warfare, Linnaeus sought a system akin to the military, with a strict hierarchy: "Nature, like society, consisted of kingdoms, provinces, districts, and individual smallholdings from which the soldier was collected." In Linnaeus's view, everything had its place in this order, with a clearly defined connection, up or down, to something else. Also called the Chain of Life, this classification system was meant to be descriptive; like a family tree, it sketched out who was related to whom, and how many degrees of separation were in between.

Its fundamental unit was the individual species. This may seem obvious, but defining "species" can be difficult. (For that matter, try to get any two random biologists to agree on a definition of "life.") "The question of exactly what constitutes a species remains among the most contentious in biology, thoroughly resistant to consensus," wrote the authors of *Vietnam, A Natural History.* Despite this, most biologists agree that species are real units defined by measurable differences in characteristics such as color, dimension, and behavior, with limited genetic exchange with outsiders.

When confronted with a brand-new, unknown specimen such as Indochina's "ferret badger" or "raccoon dog," taxonomy is much like a game of Twenty Questions. First, you ask: "Is it animal, vegetable, or mineral?"—that is, which Kingdom it belongs to—and then go from there, in increasingly specific levels.

As you go along, you get closer and closer to the subject's immediate blood relatives, and then to the thing before you. By looking at shared similarities in this way, a Kingdom can be divided into a number of different phyla, a Phylum into different Classes, Classes into Orders, and so forth.

When it came to the final, indivisible step and deciding upon the ultimate name of a given plant or animal, Linnaeus discarded existing systems, which had long and unwieldy descriptive titles that varied greatly from one language to the next. For instance, under one such system, a tomato's scientific name was *Solanum caule inerme herbaceo, foliis pinnatis incisis, racemis simplicibus.*

Instead, Linnaeus settled upon simply using a last name and a first name: one indicated the genus while the second identified the individual. Taken together, the two names made up the binomial, or "two names," system, which gave relationships as well as labels. The system was much like the first and last names of humans; we know that in the case of "John Doe," "Doe" tells us what family the individual "John" belongs to.

Fair enough. But what if you liked the binomial system and "John Doe" was already taken? Or what if you preferred "José Doe"?

With these considerations in mind, Linnaeus decided that each combination of names should be unique, and that they should all be in Latin—the universal language of science—ensuring that researchers in one country would know that researchers in another country were talking about the same thing. Now a tomato would be known to researchers everywhere as *Solanum lycopersicum,* whatever the local regional name. (To sixteenth-century French-speakers, a tomato was a "love apple." Nowadays, French supermarkets label it "tomate.")

Asked about Linnaeus's legacy, Lidén quickly came up with a short list, starting with Linnaeus's travelogues, his binomial naming system, and its implied hierarchies. He also noted that Linnaeus laid the groundwork for the theory of evolution, and that Linnaeus's use of descriptive taxonomy and his rational approach gave a model for others to follow.

Most of all, Linnaeus inspired his pupils and their successors. They traveled to the East Indies, China, the Cape of Good Hope, Russia, the Arctic, the Hawaiian Islands, Venezuela, and Arabia. His "apostles" sent back plants, animals, birds, and minerals from all over the globe, along with magnificent drawings, sketches, paintings, and written descriptions. Others later took up their cause, including Joseph Banks, who accompanied Captain Cook to Australia. Banks in turn inspired a young Charles Darwin to go around the world aboard the HMS *Beagle*.

Their passion in turn inspired new generations. Nearly a century after Linnaeus's death, explorer Alfred Russel Wallace wrote of capturing a Malaysian butterfly: "I had seen similar insects in cabinets at home, but it is quite another thing to capture such oneself—to feel it struggling between one's fingers, and to gaze upon its fresh and living beauty, a bright gem shirring out amid the silent gloom of a dark and tangled forest."

So many died in the attempt to collect new species that some scientists suggested creating a "Naturalists' Wall of the Dead" to commemorate them.

These naturalists continue to fill in the gaps about the "simple plan of nature," traveling to remote, isolated places. These days, the intellectual descendants of Linnaeus's apostles go to formerly hostile regions that were long off-limits, such as Third World tropical countries.

These days, they go to Indochina. And they enter a race against destruction.

8

A Biological Gold Rush
in the Jungle

The vilest scramble for loot that ever disfigured
the history of human conscience.

—Joseph Conrad, writing about the Central African ivory
rush of the nineteenth century

WHENEVER TILO NADLER goes out to dinner, one of the first things he does is visit the back of the restaurant kitchen, where the deliveries come in. Similarly, when he visits any farmer's market to shop for groceries, he makes sure to visit all the stalls and shops on the side streets, paying special attention to the trophies hanging on the walls. Surprisingly often, the antlers used as hat racks or as parts of ceremonial altars have turned out to be species new to science. From past experience in Vietnam, wildlife biologists have found that they never know where an endangered species—or even a brand-new one—will pop up, so they've learned to always be alert.

Vietnamese biologist Vo Quy commented, "We discovered several species by accident, only by looking at the remains left over from people's cooking pots . . . scientists like to think that they have discovered all the animals, but the everyday people see things that we don't know of."

Tantalizing hints of more new treasures have come to light in myriad pots, pans, markets, and traditional medicine stores. "Vietnam is the place to be since all the discoveries," said taxonomist and biologist Colin Groves. "Every new animal makes you think, What else is out there?"

But doing this type of search in the markets and streets of Vietnam can be difficult, no matter how small or large the town. Among other unexpected obstacles, it can be an overwhelming sensory experience, starting very early in the day while you're not yet fully awake; there is so much to take in that you don't know where to look. The crowing of roosters can be heard well before dawn in the middle of downtown Hanoi, sometimes mixed in with the pungent smell of fermented fish sauce, *nuoc mam,* wafting in from outdoor food stalls. In order to get from your lodgings to a place to eat breakfast, a visitor must dodge the wares of a florist arranging his goods into funeral wreaths, then tiptoe over a completely disassembled motorcycle spread out over the sidewalk in front of a mechanic's shop on Blacksmith Street, alias Tho Ren. While you try to examine the live animals being sold in the open-air market, street peddlers accost you hawking fresh baguettes while a few steps away old people practice yogalike moves in unison in the open public square at dawn, oblivious to the noise of the nearby traffic. (The days when everyone got around by bicycle are fast disappearing.)

The little shops adjoining the typical open-air market square can be a ready source of animal products as well, as I discovered while in the town of Cat Ba, in the nature reserve of Ha Long Bay. Under peer pressure, I had a tiny sip of a clear, numbing alcohol for sale, which, it turned out, came from a glass jug containing a dead monkey fetus on the bottom. Another jug held a dead bird, complete with feathers. Yet another contained several dead snakes.

In a mixture of languages, the elderly proprietor explained that each item was reputed to endow a certain power upon the drinker. Pointing to the monkey, he said, "Strong." Pointing to the bird, he said, "Smart." A female friend asked what snake wine was for, and the shop owner blushed. He fumbled for the right word in English. Finally, he looked around, saw no one was looking, and then pointed to his groin while slowly raising his finger upright from a horizontal to a vertical position—it was the Southeast Asian version of Viagra.

If dealing with the contents of the market stalls and shops was not enough to make one's head spin, there was the chaos of the street. Motorcycles sped past, each loaded with cargo precariously tied to bumpers and exposed to the open air. You see yellow-haired puppies in wicker baskets bound for the meat market, live ducks tied upside-down to handlebars, caged birds, bloody raw cattle ribs, trussed-up piglets, and a load of clear plastic bags full of water containing live, swimming goldfish. Once, I even spotted a six-foot-long dead shark bound to the back of a Russian-made Minsk motorcycle. To spot any new or endangered animals within this chaos is an art form; trolling the city streets and markets like this is not the first thing that comes to mind with the phrase "wildlife biology."

But it has worked quite well for some. Probably the most successful example occurred when US biologist Shanthini Dawson visited a remote Vietnamese village in what was then Nghe Tinh province, near the Laos border, more than a decade ago. An elephant researcher by training, she noticed that the village's inhabitants, composed of T'ai tribal hunters, had cooking pots full of the bones and antlers of a mammal species she'd never seen before. They were obviously not from an elephant but didn't fit anything else, either.

The body parts looked like those of a barking deer known as the common muntjac, also called the red muntjac (*Muntiacus muntjak*).

But it was of unusually large size—about a hundred pounds, or 50 percent larger than usual—and with long, curving canine teeth. DNA tissue analysis of the remains established that the animals were previously unknown to Western science. It was a new genus as well, although taxonomists had to ponder that issue for a while. (Scientists since discovered at least eight different muntjac species in the region, and there may be more. Among other complications, a previously collected muntjac from 1929 that had been sitting in Chicago's Field Museum of Natural History turned out to have been misidentified, and was actually a brand-new muntjac species.)

Now formally categorized as the large-antlered muntjac or giant muntjac (*Muntjac vuquangensis*), the animal that Dawson saw was the second new species to be discovered in the Vu Quang Nature Reserve, southwest of the city of Vinh. A new, smaller muntjac now known as Roosevelt's muntjac (*Muntjac rooseveltorum*) was rediscovered in 1994, in similar circumstances.

Vietnam's reputation as possibly the last frontier of mammalian zoology dates back to 1992. In that year a team of biologists—led by Oxford University zoologist John MacKinnon of the Asian Bureau of Conservation in Hong Kong, and Do Tuoc, a field biologist with Vietnam's Ministry of Forestry—surveyed a tract of the remote Vu Quang forest in the mountains of the Annamese Cordillera along Vietnam's border with Laos. In an isolated village, the biologists stumbled across some hunting trophies in the home of a local man. The horns of one animal—smooth, black, and spindly with a slight bow—were quite unlike anything known in that part of Asia.

That was evidence enough for MacKinnon and Tuoc to identify them as belonging to a previously unidentified species, an animal of up to 220 pounds that local T'ai tribesmen called the *saola* (pronounced SOW-la), a combination of two T'ai words that mean "spindle" and "post," referring to the shape of the horn. Four subsequent expeditions turned up twenty partial remains of the saola, known to science as *Pseudoryx nghetinhensis*.

Then, in 1994, the first live saola came into the hands of zoologists. Described by some as "so strikingly unique that biologists at first found it difficult to pinpoint its closest relatives," the animal stood a good three feet tall at the shoulders and weighed more than two hundred pounds—the world's largest land-dweller found by Western science since 1937. (Although local people certainly knew about the creature for centuries, considering that archaeologists found two-thousand-year-old jade earrings from the now-vanished Sa Huynh culture that depict what look like saola.) The animal's coloring was chocolate brown, with striking white stripes along its flanks and face. The dark color and sturdy, goatlike features of the saola would serve it well as it browses at night in patches of wet, old-growth forest. But the saola is no goat: DNA analysis showed that the animal belongs to a new genus of bovine, the group that includes cattle and antelope. The shy saola is actually a strange ox, as reflected in some other names that have been used for it: the Vu Quang ox or pseudoryx.

The discovery of the saola had biologists packing their bags and booking flights for Hanoi, and the swarm of investigators that descended on Indochina soon turned up more previously unknown species. In addition to the various kinds of muntjacs and the saola, a striped rabbit carcass turned up in a market just across the Laotian border in 1996; live specimens of what scientists came to call the Annamite striped rabbit (*Nesolagus timminsi*) were later photographed in Vietnam's Pu Mat Nature Reserve, along the Laotian border just north of Vu Quang.

And a set of unearthly horns, seemingly out of some mythical forest—with graceful, ridged spirals some eighteen inches long, and a curl at the tip—turned up in a traditional medicine shop in Ho Chi Minh City. The horns may have belonged to a *linh duong*, Vietnamese for "mountain goat," but no one was really certain. Another example of it was reputedly captured in 1929; its bones sat unstudied in a box in a Kansas museum until the 1980s. The veracity of both sets of horns was finally disproven recently, after tests with an electron micrograph showed that the graceful horns were ingenious counterfeits made for the tourist trade, with the aid of ordinary goat skulls and a pair of hot pliers over a fire. (There is a lively business in creating such novelties, which greatly complicates biologists' work. An ancient oral tradition holds that the horns of a mythical snake-eating ox will protect one against snakebite or treat the venom if bitten. Craftsmen are only too happy to fabricate whatever the market wants. "A lot of people were disappointed," said Colin Groves, the taxonomist at the ANU. "But at least nobody wasted any time and money on a special expedition.")

Even with these mishaps, the rapid-fire series of discoveries—a record fifteen new large animals discovered in Vietnam in ten years—sparked something of a biological frenzy, albeit one little known to the general public in the West and to most urban or expatriate Vietnamese.

Vietnamese biologists worried that as a result, the forests may be stripped clean of their rarest species. "As soon as you declare something rare," one told me at his office in Hanoi, "the rush begins." Westerners—including some biologists—have offered large rewards for live specimens. In the '90s, one Danish journalist offered a million-dollar bounty for a live saola; in the resulting frenzy of trapping, several of these extremely rare animals were accidentally killed. The only reason that the giant muntjac, or large barking deer, was spared was that the only specimen in captivity lived in the personal menagerie of a Laotian general, whose compound was just across the border from Vietnam. His valuable prize was guarded by soldiers toting automatic rifles.

Of this Gen. Cheng Sanyavog, *The Washington Post* wrote:

The general is famous as a nature lover. He had a pond constructed behind his house so exotic ducks and swans would be visible from his bedroom window, and has a zoo nearby.

A helpful man carrying an AK-47 machine gun showed me behind the grounds to what was indeed the general's private menagerie. There, tucked in an inconspicuous cage behind a neurotic Asian bear and a jungle gym for the kids, was the world's only known specimen of the barking deer.

The general was also the CEO of a timber-harvesting and real-estate outfit called the Mountainous Development Company, which cut down large swaths of pristine, biologically diverse old-growth forest to make way for a 268-square-mile "inundation zone" for the Nam Theun 2 Hydroelectric Project, removing thousands of native tribespeople in the process. And not just any forest, but an area of eastern Laos on the border with Vietnam that was home to some of the world's tallest, straightest evergreens, which provide habitat for muntjacs, tigers, bears, clouded leopards, striped rabbits, Heude's pigs, and the saola. Described as "a kinder, gentler dam" by its promoters, the World Bank–funded project was completed in March 2010 and supplies power to neighboring Thailand. It is the largest such project in Laos, but very little of the electricity it generates is available there.

Cutting down the forest to make way for the dam was extremely profitable, with some individual trees going for as much as an estimated $2,400 apiece, thus explaining the opulence of some mansions.

It was emblematic of what was going on just across the border from Vietnam, in the weaker, poorer, less-developed states of Indochina. "Vietnam has a reputation for protecting its own wildlife and natural environment, and then going everywhere else and hunting it out—especially in Laos," remarked George Schaller.

Tilo Nadler showed me photographs of endless rows of lumber trucks loaded with tropical hardwood lined up one behind the other at the Cambodian border, on their way to furniture factories near Da Nang and Ho Chi Minh City. The long lines were reminiscent of old photos of trucks lined up along the Ho Chi Minh Trail—which ran through the same region.

The inner workings of this head-over-heels growth are not always pretty, especially if one looks into, say, the furniture-export trade, one of the country's top five export earners. (The United States is by far Vietnam's biggest market for furniture, almost three times larger than the next-largest market, Japan.)

Vietnam is already undercutting China and taking jobs away from the Chinese, as a result of Vietnam having even weaker unions and lower wages than China, along with fewer labor laws, heavier subsidies, and bigger tax breaks to favored companies. Consequently, furniture manufacturers in China are already moving their operations from the industrial cities near Hong Kong, where workers earn $170 per month, to Vietnam—where workers doing the same labor make $80 per month. (Their American counterparts make a hardly opulent $12.00 per hour, or just under $2,100 per month— a level that Harley Shaiken, a University of California, Berkeley, labor professor, described as "only slightly above eligibility for food stamps if you have a family." Imports from Asia now make up 70 percent of the American furniture market, a 4,000 percent increase in less than ten years.)

While the mills are in Vietnam, about 80 percent of the wood itself comes from neighboring Laos and Cambodia, which have much smaller populations and less economic or military clout. Laos is about three-quarters the geographical size of Vietnam but with about one-sixteenth the number of people. Both Laos and Cambodia suffered multiple invasions from their bigger neighbors over the centuries. Much of the timber is cut in protected reserves in these countries—where laws are weak and enforcement minimal—and illegally smuggled across the border in spite of export controls, said a recent twenty-four-page undercover investigation by the London-based Environmental Investigation Agency (EIA).

Massive deforestation of the Mekong River region has been the result. The organization's field investigators made secret films during undercover visits to furniture factories and found that "criminal networks have now shifted their attention to looting the vanishing forests of Laos."

But because the furniture-export trade is worth some $2.4 billion annually, authorities tend to turn a blind eye. Corruption, large and small, has accompanied the boom times, and its reach is everywhere; I myself saw our bus driver casually insert some currency into the pages of a rolled-up magazine and hand it all over to a policeman at a routine traffic stop on the superhighway.

Pinning down exactly what is happening, and who are the good guys and who are the bad guys, can get murky. Before an interview with a high-level official, a field researcher told me to pay attention to the elaborate carved, ornamental chairs made of precious woods that adorned the waiting room; the researcher pointed out how difficult it would be for the official to afford such things on his salary—and also how this same person was in charge of breeding rare flowers on government property, ostensibly to raise funds for conservation.

Nailing it down further was tricky, as I learned from a Western official of a well-known environmental organization in Hanoi, who was a fluent Vietnamese speaker and had been living there for a number of years. "The government has very little authority out there in the hill country," he confided. Getting to the point, he said, "You can get all the official permits and everything else you want in Hanoi, but in the field you're at the mercy of the whims of the local officials. Sometimes you're at the mercy of a bunch of thugs."

He went on to explain that activities approved by the government in the city sometimes carry no weight in the field, where the local village headman reigns supreme in what is still in many ways a traditional, rural, village-centered Asian country. Family connections, friends, and local politics count for more than the actions of some official in the distant capital. It's an echo of a complaint often lodged by US military advisers about their Vietnamese allies during the war years: you can read about orders for equipment, material, and raids that were authorized in Saigon but somehow never made it to the bush.

Nevertheless, the Laotian general had his supporters, not least because he was smart enough to sprinkle around a small portion of his profits in the form of highly visible public works for the local communities, including schools, jobs, and even a Ferris wheel. The largesse gave him some political clout; *Washington Post* reporter Doug Fine said that covering the general "made me feel like I was covering a philanthropic entrepreneur in Cali or Medellín."

As a result of the economic good times, a sixty-eight-story office skyscraper, the Bitexco Tower, now crowns the skyline of the former Saigon in the south of Vietnam, whose boulevards are crammed with cars and motorcycles. The boom was even apparent in more leisurely Hanoi, formerly the capital of America's Communist enemies in the north.

At the same time as this wholesale deforestation, there was a surge in demand for wild animals. In addition to the usual trophy hunters and professional collectors, there was a newfound taste for expensive exotic game, an unfounded belief in the curative powers of medicines made from wild animals, and a desire to make a fast buck. Taken together, these made for a huge illegal, black-market trade that is as much of a threat to wildlife as loss of habitat or the expanding human population. The estimated global value of this trade could be as much as $20 billion annually, wrote the US Congressional Research Service.

This market has an appetite for everything imaginable: horns, bones, bladders, paws, tails, and fins.

No creature is seemingly too insignificant to be desirable to someone for something. Biologist Mike Hill of Fauna and Flora International of Cambridge, England, recounted one example: the plight of a species of long-horned beetle, a large and attractive insect

that can take years to mature. A well-known "ecological island" of
the beetles lies on the outskirts of Hanoi, and outsiders are not allowed to gather them there. But local people are exempt from the
restrictions, and so outsiders would simply buy collection permits
for the locals and purchase their catch. Large quantities of the rare
creatures went for hundreds of dollars.

Ecologists say that the forests of Southeast Asia, a biological
treasure trove, have become a gold mine for wildlife traffickers,
with Vietnam as the major Asian crossroads for trans-shipment
from poorer countries in the region to wealthier ones. Animals are
smuggled to Vietnam from Laos, Cambodia, Indonesia, and Burma
for export to markets in Taiwan, China, South Korea, and Hong
Kong. They include macaques from the Mekong Delta, gibbons
from the Hanoi region, and Asian black bears and sun bears from
the Vietnam/Laos border.

Vietnam banned hunting without a permit in 1975 and has
signed several wildlife protection treaties, including the Convention on International Trade in Endangered Species of Wild Fauna
and Flora. Yet enforcement is often weak, and the estimated profit
of the illegal wildlife trade is thirty times larger than state spending
to combat it.

Accompanying these factors was a loosening of government
restrictions and a booming trade in wildlife smuggling across the
Chinese border to satisfy that country's nouveau riche; one Hanoi
University expert estimated that three thousand tons of wildlife
and wildlife products are smuggled out of Vietnam every year, with
only about 3 percent intercepted. The animals are eaten as status
symbols by the newly wealthy, their bones ground up for use in traditional medicine, their skins and heads hung on walls as trophies.

Consequently, many of Vietnam's wild areas are in danger of
becoming denuded of animal life, creating what organizations such
as the Switzerland-based World Wildlife Fund called "empty forest

syndrome." More than three hundred animal species have disappeared in Indochina; more than one hundred others are threatened.
With virgin rain forests steadily reduced, possibly as few as between
forty and one hundred wild elephants are believed to survive in
Vietnam's wilds. (To put this in context, as recently as 1980, there
were an estimated 1,500 to 2,000 elephants in Vietnam. And the
traditional name for Laos was "The Kingdom of One Million
Elephants.")

And what happens here in Indochina is a template for the rest
of the tropics, where the most biological diversity is located. This
voracious pattern is especially true for closed societies expected
to open up someday, such as Myanmar (Burma) or Fidel Castro's
Cuba. The latter is home to the world's tiniest bird and the smallest frog, along with intact tropical ecosystems and heaps of rare
and endemic species; Cuba's Jardines de la Reina reefs are said to
still be "just as Columbus saw them." Embargoes and the fumbling
economic policies of Cuba's government have kept it all from being
developed, but that may end soon. Already, organizations such as
the Society of Environmental Journalists sponsor weeklong tours of
Cuba, with the slogan "Cuba Opens Its Doors . . . To What? Visit
Paradise Before It's Paved." (The US embargo makes exceptions for
American journalists and educators to visit Cuba, as part of cultural
exchanges.)

As economies and societies open up, mass-market development
moves in and habitat becomes lost. According to Harvard biologist
E. O. Wilson, three species of plants and animals are lost every hour
in the rain forest, or thirty thousand species per year. It's the greatest
mass extinction since the comet that killed off the dinosaurs.

At the same time, there is some cause for hope. Between 1997
and 2007, at least 1,068 new species have been discovered in the
Greater Mekong—the river system that drains Indochina—for an
average of two new species a week every year for the past ten years.

And some of these creatures are big: besides the saola and the munt-jac, there were two-hundred-pound turtles, fifteen-foot-wide pad-dlefish, and more.

The result for these newfound species seemed to be a sort of race between the forces of preservation and those of extinction, in which everything seemed to be up for grabs and nothing was too outrageous. Even the contents of the animals' bladders are actively sought, sometimes by horrific methods.

I learned more while visiting the headquarters of the *Vietnam News,* Vietnam's national English-language daily, one of whose editors, Howard Smith, I met by accident while in a Hanoi coffee shop. The Australian-born Smith invited me up to the paper's of-fices. I had no idea what to expect inside the anonymous building.

Smith led me to an inconspicuous door in a quiet hallway; opening it revealed a busy newsroom, full of dozens of reporters making phone calls and typing away at laptops. The roar of activ-ity made me feel right at home—as well as a little homesick for the bustling newsrooms I knew in college. Showing me around to some of the dozens of staff, I saw that they were a mix of Vietnamese and Westerners, trading information and aggressively tracking down leads while talking shop. Considering that the publication was only quasi-independent, the din of activity felt just like any paper back home. "The reporters know which way the wind is blowing inside the government, and they have a good sense of how far they can push things," one subeditor, a Westerner, told me.

Hearing of my research, he dug up a story that one of the Viet-namese staffers had done, entitled "Bears Tortured to Meet Asian Thirst for Bile." It described how more than four thousand living individuals of a species known locally as "wild moon bears" were being kept in small, five-by-five-by-eight cages in city centers, so that the contents of their gallbladders could be periodically drained. The animals, restrained with ropes, often without anesthetic, have their abdomens "repeatedly jabbed with 4-inch needles until the gall bladder is found, and then the painful extraction begins."

The reporter, Quynh Anh, didn't put it in the published story, but said that he could hear the animals' cries from the street outside the building.

Because each ounce of fluid is worth about $100—and there are several ounces extracted at a time—the trade is very lucrative. Education for Nature Vietnam (ENV) says that the bile-farming industry has exploded in numbers over the past ten years, from the low hundreds to the thousands.

Part of the newspaper story read:

Many Asians believe bear bile is capable of curing even cancer. That's why many people are ready to pay a lot of money to buy it. However, it is doubtful if bile taken from such mistreated bears has any medical benefits, not to say possible harmful effects, according to doctor Dang Van Duong, chief pathologist at the Bach Mai Hospital in Ha Noi.

After conducting clinical examinations of the damaged gall bladders of three rescued bears at Tam Dao sanctuary, he concluded: "Had those gall bladders belonged to humans, they would already be dead."

According to Animal Asia Foundation, many traditional medicine doctors agree that bile can be substituted with 54 recognized herbs. Ursodeooxycholic, the main ingredient in bear bile, can also be synthesized under laboratory conditions. Both herbal and synthetic options are readily available, making the use of bear bile unnecessary.

It was shocking and surprising, but other undercover reporters had encountered the same thing, such as a years-long investigation for *National Geographic*. The story was illustrated with a photo that showed a sedated Asian black bear lying spread-eagle on its back on the floor of a facility in Vietnam, hooked up via intravenous tube to a number of medical devices. Its bile can be seen dripping into an adjacent bottle.

Similarly, a Chinese animal-rights activist, Xiong Junhui, spent four years infiltrating the bile market before being able to make an undercover video of the milking, which is sometimes done three times per day. The frighteningly thin bears are kept in tiny cages, with permanent holes in their abdomens and metal jackets around their bodies to prevent movement; the animals are sometimes confined in this way for years.

The bear farms are similar to tourist attractions called tiger parks or zoos, which secretly operate as fronts in which the captive animals are butchered for their parts. (Captive animals are exempt from the CITES treaty, which only applies to wild animals. Smugglers can evade the law by establishing phony breeding centers and claiming that all their products are from captive specimens, using phony papers.) Now, it seemed, the idea of using phony breeding centers as fronts for black-marketeering had taken hold in Vietnam as well; there are at least seven "bear farms" in northeastern coastal Quang Ninh province that are selling bear products. Across the border in China, some of these farms are quite large, holding up to five hundred of these endangered animals; the owners claim that the procedure is no more painful than getting an ear pierced.

However, the practice was described as "unconscionably cruel" by the founder of the Hong Kong–based foundation Animals Asia, Jill Robinson. The foundation runs a sanctuary in Vietnam, which has rescued about ninety moon bears. In addition to saving the individual animals, its veterinarians publish their findings about the inflamed organs, contaminated bile, scar tissue, embedded objects, and tumors that they've found in the animals, in hopes of influencing public opinion. Robinson told *OnEarth* magazine's Barry Yeoman that progress has been made, with the Chinese government closing some bear farms, and that China should not be written off "as a nation of animal abusers and animal haters." The tide of public opinion in China seems to be changing, if slowly.

And in an effort to combat these and other abuses, since 2005 the Vietnamese government's Forest Protection Department has made an effort to register and microchip all captive bears.

The cases that are picked up are the tip of the iceberg, said a representative of the World Wildlife Fund (WWF). "Wildlife populations are dwindling at an alarming rate due to illegal trade and consumption," said Eric Coull, who heads the group's Greater Mekong section. "Nowhere is this more evident than in Vietnam."

According to Nadler and other researchers, the border towns see a steady, daily parade of vehicles going by, containing pangolins, lizards, cobras, pythons, monkeys, bears, and tigers, among others. The creatures have been smuggled in and out of the country via ambulances, hearses, and wedding limos. Permits and licenses are sometimes forged, wrote a Vietnamese academic, and customs officials bribed or threatened; organized crime syndicates are suspected to be involved.

As a result, illegal hunting and trading are at an all-time high, said a representative from the wildlife trade–monitoring network TRAFFIC. (The organization is a thirty-year-old joint program of the WWF and the International Union for the Conservation of Nature, working to ensure that trade in wild animals and plants will be managed at sustainable levels.) In July 2010, customs officials in Guangdong, China, seized nearly eight tons of frozen pangolins—scaly, anteater-like animals—and nearly two tons of the animals' scales from a smuggler's fishing vessel. The vessel's Malaysian crew had received instructions by satellite phone as to where to rendezvous at sea to pick up the contraband from a mother ship and transfer it to smaller boats bound for the mainland. "The use of satellite phones and trans-shipment of cargo at sea are indicative of the increasingly sophisticated methods being used by the organized criminal gangs involved in wildlife crime," said TRAFFIC's Asia-Pacific monitor.

In another twist, Vietnam's latest law about confiscated animals has a serious loophole: it says that only healthy animals must be returned to the wild. Unhealthy ones can be sold legally and the money pocketed, so there's a strong financial motive to declare all confiscated animals as "unhealthy," regardless of their true physical condition.

The sums of money involved can be large. Experts put the value of the parts of a tiger at between $10,000 and $15,000, with reports of some animals earning up to $30,000. Their heads are put on walls as trophies; their meat sold in restaurants catering to exotic fare; and their bones used for tiger-bone wine, sometimes called "the chicken soup of Chinese medicine." The trade in "crazy Chinese medicine," as Nadler once called it, was extremely profitable. Nadler said that sometimes the VIPs who visit his Endangered Primate Rescue Center expect to stock up on supplies for folk medicines. "So many well-educated people tell me 'I don't believe in traditional medicine—but when I get really sick, I try gallstones or rhino horn.'"

The demand for rhino horn in particular has skyrocketed, so much so that an organized gang of thieves recently broke into a museum in Ipswich, England, and stole the horn off a stuffed rhino that had been in the museum's collection since 1901; authorities suspect the same gang is behind a rash of such thefts all over Europe in 2011, in which fifty-eight horns were stolen from sixteen different European countries. Meanwhile, in the USA, seven rhino-horn smugglers were arrested in February 2012, in "Operation Crash," a multiagency effort headed by the Department of the Interior.

After a little piece of horn is sold to the end user, it is then ground down into powder and mixed with water or tea and drunk as a medicine in Asian countries, where it is commonly believed to cure everything from impotence to cancer. The black market in rhino horn played a direct role in the extinction of the last Javan rhino in Vietnam, judging from the fact that it was found dead with its horn hacked off. Vietnam seems to have become the major epicenter for the illicit rhino-horn trade—particularly after a prominent Vietnamese politician claimed it cured him of cancer. Nearly all black-market rhino horn seems to pass through here at some point. Sometimes it is used in Vietnam itself, but often the horn is in transit to other places, such as China.

Fetching $60,000 a kilo, or nearly a quarter-million dollars for one average-sized horn, the trade is difficult to stop; the Natural History Museum in London now puts fake horns on its rhinos, while the Natural History Museum in Bern, Switzerland, sawed off the horns of six of its rhino specimens, replacing them with wooden fakes in February 2012. And some game reserves in South Africa now remove the horns from their rhinos to make it financially worthless for poachers to kill the animals.

The ironic part is that, scientifically speaking, the horns have no therapeutic value whatsoever. One scientist, Raj Amin of the Zoological Society of London, said, "There is no evidence at all that any constituent of rhino horn has any medical property. Medically, it's the same as if you were chewing your own nails."

The fact that so much wildlife is dying for something that doesn't even work can drive some wildlife biologists crazy. "The makers of aspirin need to flood the Vietnamese market with inexpensive aspirin, with some advertising slogan that tells them that it is far, far better than rhino horn, tiger bone, or langur wine," said mammalogist Colin Groves.

But the facts didn't seem to matter.

And the reason for the booming black market included more than just quack medicine. There was something else going on as well. Like people elsewhere in the southeastern part of Asia, Vietnamese people often express pride in their adventuresome culinary tastes. A popular saying in the region goes "We can eat anything with four feet except the table. We can eat anything in the ocean except submarines. We can eat anything in the sky except planes."

Along the same lines, Tilo Nadler had told me that often, after his staff has conducted a tour of their endangered animals at the EPRC, the first thing they heard from their distinguished visitors was "We've seen your animals. Where's your restaurant? Now we want to eat exotic wild animals."

The reasons for this newly acquired taste can be complex. According to Edwin Wiek of the Wildlife Friends Foundation, a wildlife trade–monitoring network now headquartered in Thailand, the eating of rare or endangered species has become a sort of twisted status symbol in Indochina. "The fact that you can get tiger meat shows you have money. It's illegal, it's difficult to get. It's like caviar," he said to a reporter from Agence France-Presse. "For some people, having a Ferrari outside their front door is not enough; you have to have a chimpanzee in your backyard as well. Then you're really the man."

All of this is not to single out the Vietnamese; in fact, the United States is the biggest consumer of trophy species trans-shipped from Vietnam. The West is not immune to the quick-and-dirty mentality of doing anything possible to make a fast buck from a country's patrimony.

I became vividly aware of this when reporting from Australia for the old *New York Newsday,* where organized gangs were stealing fossils from national parks and Aboriginal sacred sites for the black-market trade overseas. Some of the objects were quite rare, upsetting paleontologists and anyone else interested in the country's natural history. "These tracks were the only evidence of *Thyreophorans* [four-legged, plant-eating armored dinosaurs with spikes or plates] in this part of the globe. There are less than a dozen examples of it worldwide," said paleontologist Tony Thulborn of the University of Queensland. Ken McNamara, curator of invertebrate paleontology at the Western Australia Museum, agreed: "Paleontology is like trying to put together a giant jigsaw puzzle when you've only got half a dozen of the original pieces left. Now someone's taken one of those away." The scientists were also concerned that the loss of the fossils jeopardized relations with the Aboriginal community, causing paleontologists to lose access to specific, nationally registered locations. At one point, the elders from the Goolarabooloo clan went on Australian radio to place a traditional Aboriginal curse upon whoever had stolen fossils from the Goolarabooloo sacred site.

"If you were out to make money, you ought to smuggle fossils instead of drugs," commented Thulborn. The trade is so big that some fossil hunters even advertise their latest trophies for sale on the Internet.

Thulborn and others are concerned that their country's heritage is being sold off to the highest bidder, disappearing overseas before researchers can even take a look at it. These concerns are not limited to Australia; the Manhattan branch of Sotheby's auction house sold a *Tyrannosaurus rex* excavated from South Dakota for more than $8 million, over the objections of organizations such as the Society of Vertebrate Paleontology and the Dinosaur Society.

However, an outright ban on all fossil collecting was not desirable either, say some scientists. "There is a case to be made for removing tracks for safekeeping from areas of active erosion, where fossils would otherwise be lost to the elements," pointed out John Long, curator of vertebrate fossils at the Western Australia Museum. Thulborn agreed, adding, "Much of paleontology depends upon the efforts of honest amateur collectors bringing material to museums." In an era of tight budgets, museum curators buy pieces from outsiders rather than go to the time and expense of mounting their own dig; some institutions even publish "wish lists" of items they want.

In the absence of any clear-cut rules, thieves in the land down under have grown bolder. A well-organized gang removed Ediacaran fossils from several different reserves in the Flinders Ranges

National Park. The fossils, worth tens of thousands of dollars on the commercial market, were scientifically priceless: "Ediacarans consist of fossil jellyfish, soft corals, worms, and things that have no modern counterparts. These fossils mark the first appearance of complex life-forms on the planet above the cyanobacteria level," said Ben McHenry, collection manager of earth sciences at the South Australia Museum.

The same gang used dynamite to blast trilobites out of a cliff on Kangaroo Island, South Australia, the only site of the fossil-rich Burgess Shale Formation in the Southern Hemisphere. "The trilobites, made of red calcite on a gray shale background, were really spectacular and eye-catching," McHenry remarked. "It was a scene of absolute devastation when we got there," commented his colleague Neville Pledge.

But things may be changing. The missing trilobites and ediacarans were ultimately traced by Australian Customs and Interpol, who found the gang with crates of material, some of which had already been sold and delivered to a museum in Japan. Two men pled guilty to charges of exporting without a permit, and the spectacular thefts attracted national attention, paving the way for new proposals to deal with the growing problem—such as hiring Aboriginal rangers to guard the country's natural patrimony in remote regions of the outback.

Meanwhile, Thulborn had some suggestions. He no longer publishes the exact location of a fossil find in scientific journals, for example. Instead, he writes generalities about the location and then entrusts details to the state museum or other authorities.

Most critically, he thinks better public education would also help: "People don't go into art museums and snap bits off statues or cut pieces out of paintings. It's not just because there are guards there, but because they know better."

9

The Kouprey—
A Cautionary Tale

Pol Pot killed two million people. As far as I know the kouprey never
bothered anyone. But they were both rather elusive. The kouprey
proving more so than Mr. Pot.

—Nate Thayer, the last journalist to interview Pol Pot alive
in the jungle, describing his subsequent expedition to find
the kouprey

PROBABLY THE ULTIMATE test case for showing that we should
know better, and be more careful and more diligent about protecting specimens of Indochina's natural patrimony, comes from
the tale of the kouprey—a living, breathing animal species whose
exploits have earned it near-legendary status. The creature was in
a Western zoo only once and never held in captivity again, despite
strenuous efforts by science. Its story dramatically shows the many
near misses, close calls, and almost successes that have taken place
all too often in Indochina's Wild West. It also shows how easy it
is to take a species for granted—and how hard it can be to find it

again. And the experience with the kouprey may also act as a warning of what to expect when it comes to protecting the saola and the muntjac.

It all started simply enough, in a routine and straightforward manner, with no hint of what was to come. In 1937, a veterinarian named Dr. Sauvel was called to a Paris zoo in the suburb of Vincennes to identify a mysterious young calf that had been found accidentally mixed in among a shipment of wildlife from the colonies in Indochina. The shy animal moved gracefully, and it did not resemble the cow-like banteng or the massive buffalo-like gaur, two much more common, well-known large wild animals from the region.

In addition, everyone remarked upon the new animal's unusual, lyre-shaped horns that corkscrewed upward and were completely unlike anything seen before; they were among the longest and widest horns of any of the bovine family. Sometimes the tips of these horns got frayed, from the males' habit of digging in the soil with their heads, resulting in "tassels" that make the horns even more distinctive. The creature could move deer-like at a light trot, reaching speeds of up to twenty miles per hour, or thirty-two kilometers per hour. When it grew up, the calf would have gray skin, an enormous dewlap, weigh almost two thousand pounds, and stand nearly six feet high at the shoulder.

Sauvel was well suited to determining the species because he used to live in Indochina, where he was called "one of the most famous hunters in North Cambodia" and had once examined a similar, if dead, specimen. He told the zoo authorities that what they had on their hands was a fabled wild forest ox known as a kouprey, or "koup proh" as the ethnic Khmer who lived in the animals' habitat called it. Though the animal had long figured in the region's culture—there are carved kouprey statues everywhere, including in bas-relief on the walls of Angkor Wat—it had rarely, if ever, been seen by outsiders, even though it dwelled in the relatively open dry savannahs, woodland meadows, and scattered glades that pockmark the forests where Laos, Cambodia, and Vietnam meet.

It was the first such live specimen ever held in a zoo, and in honor of the occasion, the animal was given the scientific name of *Bos sauveli,* after the good doctor.

Little did those present know that this would be the one and only authenticated specimen ever kept in a zoo. (Although decades later, the leader of Cambodia, Prince Norodom Sihanouk, did once claim to have had a calf on the palace grounds when he was a child, and the animal has since become the national symbol of Cambodia.)

After electrifying the scientific world with its presence, the only known kouprey in captivity disappeared sometime during the hard times of the German occupation of Paris in World War II. No one knows what happened. The calf may have starved to death or been eaten. It may even have wandered off during the tumultuous days of the city's liberation by the Allies in 1944; in the book *Is Paris Burning?* by Larry Collins and Dominique Lu Pierre, there is an account of zoo animals and circus escapees roaming the streets of Paris after a punitive German aerial bombing raid. All we know for certain is that a single kouprey individual was discovered by Western science—and that none is known to have been held in a zoo again, despite some near-successful attempts to capture a specimen half a century ago. The animal is so elusive that it has rarely even been seen, and has been written off as extinct several times.

So when Hanoi biologist Le Vu Khoi saw a kouprey in 1990 in the region of Duc Lap, Vietnam, it set off a wave of excitement, even though he had been unable to get any photos of the animal. This was the first time a kouprey had been definitively recorded since zoologist Charles Wharton's ill-fated expeditions of the 1950s, which took place just before the series of clashes that closed off the whole region for decades.

The reason for all the excitement and effort about the creature lies rooted in the fact that the kouprey is probably the most primitive of all living cattle, containing features very similar to its 600,000-year-old forebears—much like another wild ancestor of cattle, the zebu. These features place it close to the trunk of the Linnaean tree of life, making it a good candidate for domestication and for breeding stock, says Noel Vietmeyer, a specialist in the economic value of tropical species at the National Academy of Sciences in Washington, DC. In fact, Vietmeyer wrote that he suspects that the animal may have been domesticated briefly during the height of the Khmer empire, between four hundred and eight hundred years ago. The animal could be important to studies of the evolution of cattle, and a genetic resource for crossbreeding.

Altogether, these facts mean that the genes of a single kouprey could be worth billions of dollars to industry, with a new, genetically improved "supercow" as the result. Crossbreeding one of these seldom-seen, rare wild animals with domestic cattle could offer a huge genetic boost to domestic cattle bloodlines in terms of disease resistance and hybrid vigor. Any offspring would likely be immune to such common cattle diseases as rinderpest or hoof-and-mouth. Consequently, Vietmeyer described finding a live one as "the Holy Grail" of the field, explaining: "Here's an animal with thousands of years of survivability in the harshest habitats built into it, one that could improve the lot of half the domestic cattle on Earth . . . and it's only a gleam in our eye because no scientist has seen this thing up close in forty years."

The potential value is so high that the Vietnamese government has gone on record as saying that even if a kouprey were found in Cambodia, its genes would belong to Vietnam.

But getting one has not been easy.

One Hanoi biologist, Ha Dinh Duc, was following the trail of a live kouprey in 1990 when he was shot off the elephant he was riding and nearly killed by a stray band of die-hard Khmer Rouge rebels while on the Cambodian side of the border; he was grazed in the chin and chest, and his wristwatch was broken by a bullet, while three other people were wounded. When he went back a year later, he caught cerebral malaria.

Members of another expedition also came down with malaria; a participant in a third search led by a Thai doctor named Boonsong Lekagul was severely injured after stepping on a land mine; the list goes on and on. Some wonder if the animal still exists, or if it was all imaginary in the first place, much like the Nguoi Rung, which is sort of the Bigfoot of Indochina.

But then again, there was that kouprey calf in the Paris zoo . . . along with a few documented photos from the wild in the early 1950s, and a sighting in the late 1960s by a conservationist named Pierre Pfeffer of the World Wildlife Fund.

"It's amazing the bad luck, the problems, the frustrations," said Charles Wharton, now deceased, one of the few to photograph this "missing link" between modern domestic cattle and their more primitive ancient ancestors.

In the early 1950s, before full-fledged, total warfare came to Cambodia, Wharton led a ninety-man expedition—of whom sixty were armed soldiers for protection—to look for the elusive creatures in some of the country's more remote provinces. At the time, he estimated that there were about six hundred to eight hundred of the animals. The team managed to capture five kouprey, but two died and three escaped, in a series of accidents. Just when the team was getting close to success, they were warned to get out of Cambodia as soon as possible, because the Vietnamese were about to overrun the region in the lead-up to the finale of their war against the French. No kouprey have been caught since, but there are still some disputed sightings. A few minutes of Wharton's rare footage of a distant herd of kouprey in the wild are all that we have, which can sometimes be seen online and in television documentaries today.

Hendrix, author of the subsequent Wharton article, "Quest for the Kouprey," wrote: "The most painful [experience] of all has been the excruciating near-successes of fresh tracks, second-hand reports and botched captures. To show for it all, science has amassed a kouprey collection amounting to little more than a couple hundred pounds of bones and a few feet of grainy film footage."

Because of the scanty evidence, some scientists even question whether the kouprey is a true species at all, or whether it is the result of a common banteng, or wild cow, mating with a domestic cow that had wandered off into the woods. Researchers Hunter Weiler, an adviser to the Cambodian Wildlife Protection Office, and Gary Galbreath and F. C. Mordacq wrote in the *Journal of Zoology* that any resulting progeny would have been a "feral hybrid," living in the forest, and reverting to a wild state. Using DNA analysis of kouprey and banteng remains in 2006, they concluded, "The kouprey was not a natural species." Other scientists argue that the kouprey had once been a distinct species but began to interbreed with the more common banteng as its own population declined and became fragmented; still others think the kouprey could have been a descendant of current domestic zebu cattle.

Most biologists are still waiting for more evidence. Some, such as Nicholas Wilkinson of the Cambridge University Darwin Initiative Project, think that there is a strong likelihood that there's a remnant kouprey population somewhere, but so small and scattered as to be functionally extinct.

In short, some researchers think it is out there; some think it isn't; some think it never truly existed as a distinct species; and some think it may be there but not as a viable population—a position held by zoologists George Schaller and Alan Rabinowitz. In addition, some think the animal isn't worth the huge amount of time and money that would be required to pull it back from the brink; Vietnamese biologist Le Khac Quyet pointed out that "the last few, surviving members of any species are going to be incredibly wary. They have to be, or else they would not have gotten that far."

The "it's-not-worth-the-effort" argument is something that Wilkinson, for one, has trouble with. He said, "Even a one-percent chance of success is worthwhile. You take what you can get; and one percent is better than one half of one percent. . . . Besides, I'm just not comfortable about picking and choosing which species will survive and which will not. It's like playing God. And it also raises the question of who's doing the choosing, and what their priorities are. If there is not a general, community-wide consensus, then what does it mean when one person says, 'I'm not going to bother with this species'? It seems better to just try a blanket approach, and save all you can."

The search for the kouprey is much like the current effort to save the saola—an animal that became sort of the poster child for Vietnam's newly found wildlife. About the size of an antelope, with long, backward-curving horns, a brownish color with white stripes on its face, and dark legs, the saola caught the world's imagination in 1994 when it was first discovered in the dark, wet forest of the Central Highlands—the first new large mammal to have been found anywhere in the world in the last half century. But after it was discovered, the frenzy to trap one to satisfy collectors caused its numbers to crash from about a thousand to a couple hundred.

Bill Robichaud, coordinator of the Saola Working Group, wrote, "We are at a point in history when we still have a small but rapidly closing window of opportunity to conserve this extraordinary animal. That window has probably already closed for another species of wild cattle, the kouprey, and experts at this meeting are determined that the saola not be next."

Finding more examples of it has been problematic these days. Wilkinson said, "We have been using camera traps to look for the saola and two muntjac species that have a similar range and are only found in this region. We've had zero success. The saola roams a lot—an area of about four to five square kilometers. And it is incredibly shy, stepping so quietly and softly that villagers call it 'the polite animal.'"

His colleague, Ben Rawson of Conservation International, felt that there was a mistaken sense that we should keep looking for more samples of it at all costs, over and above all else. "Everyone said that we don't know enough about the saola and we need to capture it in order to find out more. Well, we've been trying that for years," decried Rawson. He added, "Maybe we should focus more on the threats and the pressures that this animal faces, such as habitat loss or hunting. Putting up more camera traps does not prove anything, other than that it exists. We already have photos of it. The money it takes to make absolutely sure about its numbers in a given locale is huge; for the expense and effort that it takes, you could have removed all the snares in the area, which is a much more important job."

Nevertheless, people keep trying to catch a live kouprey. "All the real wildlife is in the border areas and other margins; that's where the action is," Ben Hodgdon of the World Wildlife Fund said with a shrug during an interview at WWF headquarters in Vietnam.

Probably the most audacious attempt to catch a kouprey during the postwar period has been that of correspondent Nate Thayer, whose previous exploits include finding and interviewing Pol Pot at Khmer Rouge headquarters—and coming back alive to tell the tale. Thayer says he first heard about the kouprey after some astonishing circumstances: he ate part of one during a meal deep in the forests of northern Cambodia during a two-month, seven-hundred-kilometer trek with a renegade guerrilla army. However, at the time, all he knew was that his companions had shot a large jungle bovine that they called "koup proh" and that he was having some of the meat for dinner.

And that his hosts claimed that they saw the creatures all the time, and that they considered the animal to have mystical powers.

Upon getting back to Phnom Penh, he discovered the significance of his dining experience. He immediately did several aerial and ground surveys, and then organized an epic twenty-six-person, six-elephant expedition to find the kouprey, with the approval of the king and the prime minister. "It was more Wild West in those days, and you could wangle your way into anything if you knew the tricks," Thayer wryly explained.

The experience was hairier than his experience in finding the leader of the Khmer Rouge: the plane crashed, the security team mutinied, people collapsed from exhaustion, and some thought they were all going to die when they encountered armed guerrillas.

However, Thayer added that this was all part of the appeal. "For me, that was the 'art' in it all."

When asked if he would search for the kouprey again, Thayer told me: "I spent fifteen years in that country. I did everything I wanted to do. I found Pol Pot. I chronicled the war. Plus I was expelled three times, blown up by a land mine, had eight of my personal bodyguards killed, had three assassination attempts, was taken hostage twice, had malaria sixteen times. I don't really want to push it much farther. . . . The current government is not a fan of mine. Plus the traditional environmental organizations really didn't like us doing this kouprey search and tried to block it because it undermined their not-very-creative fund-raising approaches. . . . So the answer is no . . . I have no interest in doing it again."

I later learned that Ha Dinh Duc—the researcher who was shot off the back of his elephant—had decided to switch his focus from kouprey to nothing more strenuous than the turtles that live in the lake in the middle of downtown Hanoi, an effort at which he succeeded so well that he became known as "the turtle professor."

10

Cobras Under the
Kitchen Sink

The decrees of the Emperor end at the village gate.

—Traditional Vietnamese saying

IF YOU EVER want to see what biodiversity looks like, just leave
out some laundry on a clothesline in the tropics at night.

A group of about a dozen people—including me; some EPRC
volunteers; a few primatology grad students; Tilo Nadler; his wife,
Hien; and their children, Khiem and Heinrich—are having dinner
at the Nadlers' home.

At the start of this spring 2010 evening, Hien had hung up some
sheets to dry on the deck of their stylish, modern Western-style
house just before dinner. Located in the middle of the grounds of the
EPRC, where they can keep a close eye on the langurs, the house
tonight is the site of a much-anticipated home-cooked feast prepared by Hien. This is a good chance for visitors to enjoy some genuine, real Vietnamese cuisine—no "spaghetti" made from a packet

of instant ramen noodles and ketchup, which I had suffered through once at the equivalent of a truck stop outside Hué.

Tonight, we have a chance to taste a variety of foods, which I dutifully try to keep track of in my journal. Before being overwhelmed, I manage to record: fried spring rolls filled with shredded pork shoulder (*Nêm*); tangy noodle soup with beef (*Pho bo*); sweet mint; spring onions; rice with fermented fish sauce (*Nam muoc*) that itself is flavored with garlic, sugar, lemon juice, and sliced red chili peppers; lemongrass; long, thin, narrow, rolled-up Vietnamese pancakes (*Banh cuon*); jackfruit; and a dessert made of lotus seeds and a sweet, lychee-like native fruit known in the West as a longan. The meal was followed by several near-mandatory shots of Vietnamese brandy, which mellowed out the crowd considerably and loosened up some tongues.

It was a good chance for everyone to talk shop about conservation work. And a good opportunity for me to try to figure out just why some rescue projects—such as the Nadlers'—are successful while others are not.

But first, we have to attend to the sheet, which the children discovered was a wonderful attractant to the local nightlife. When the sheet was hung up, it was inadvertently placed in front of a bright outdoor porch light; as the multi-course meal wore on and night came on, more and more moths were attracted to the bright glowing white sheet, located just outside the sliding glass door of the dining room. By dessert, the plain sheet was nearly covered with fluttering moths and other insects of the night, some of which look like scintillating colored jewels. There are insects in brilliant metallic reds, golds, and greens, along with colors whose names I'd have to look up in an art book.

Seven-year-old Khiem and four-year-old Heinrich are delighted by the sight, and make crayon drawings of what they see. The adults become fascinated too, and several dinner guests bring out cameras, flashes, and close-up lenses. (Field biologists seem to be devoted adherents of the Boy Scout motto: "Be Prepared." I've learned not to be surprised that they have gear at the ready.)

Most of the insects are moths, which I had always thought of as just dull, drab, boring, gray objects. But the ones in the tropics are stunning, coming in all shapes and sizes. One large moth we see tonight is a delicate shade of palest blue, with four little pink spots. Another looks to my eye like a miniature stealth bomber, only in pure white with gold trim on the edges of its wings.

Moths, it turns out, have their passionate admirers world-wide, ranging from Vladimir Nabokov, author of *Lolita,* to famed nineteenth-century traveler and writer Margaret Fountaine, subject of *Love Among the Butterflies.* There are identification manuals to moths and all their subcategories; one publication is memorably entitled *A Field Guide to Bird-Dropping Mimics.* There are also commercially available light traps, an annual "National Moth Night" in Britain, websites devoted to the latest moth sightings, at least one magazine, and a blog, *Martin's Moths,* which carries the subtitle: "A Fascinating Insight into the Secret Lives of Moths and Men."

But from what I can gather, it seems that nowhere are moths as spectacular as in the rain forests of Indochina. We all marvel at what we see on a piece of simple white bed linen at night. Tilo and the others start rapidly rattling off moth names, such as "tiger moths," "clearwings," "grass moths," "leaf-hoppers," "plant-hoppers," and light-emitting "net-wing beetles." That's nothing, says one guest, who tells us that the creatures once covered a sheet in Laos so thoroughly that no one could even see the fabric underneath.

It's a stunning testament to the wonders of biodiversity, and how even the lowliest creatures can take on new dimensions once you start looking closely enough.

Of course, it's one thing to appreciate biodiversity but another to protect it, which is the concern of those here. Tilo Nadler is probably in the best position to describe the situation in Vietnam and explain why some Western-initiated protection efforts succeed and others fail; with nearly twenty years of field experience in Vietnam, he is sort of our interpreter between East and West.

In fact, everything in this house relates back to the wilderness of Vietnam. The large and spacious two-story structure is full of a mix of Oriental and Occidental: the walls are mounted with animal heads and wooden crossbows, and the lampshades are made from the traditional conical straw hats worn by local peasants. At the same time, the house has all the latest modern Western appliances, including a brushed-aluminum refrigerator and a microwave.

The family flies to Germany about once a year to visit Tilo's relatives; there's Asian art scattered about; and the living room has children's schoolbooks in Vietnamese, English, and German, next to stacks of monographs, reports, and scientific papers that Nadler has written. Everything seems to have a tie to the environment of Indochina; the very table that we are eating upon is made from a single six-inch-thick, four-foot-wide, fifteen-foot-long slab of rich tropical hardwood that was seized from a timber smuggler. Once a tree like this has been cut down and sawn up into pieces, there's not much you can do. You cannot very well return it to the wild, and it seemed wasteful to throw it out, so the unfinished slab wound up here.

The objects also speak to the rapid growth of the EPRC.

The land area of the semiwild enclosures has tripled in size, and where there used to be a few full-time leaf cutters, there are now seven cutters from local villages gathering three times as much as before. (By hiring local people, the center is also getting money into the hands of villagers who might otherwise be forced to resort to poaching to survive. In a sense, it is giving the villagers an economic stake in the success of animal protection.) There are now forty large and spacious "enclosures" as well as the two large outdoor "training facilities" where the animals can run around much like sheep in a farmer's fenced pasture. The list of sponsors runs as long as my arm; besides the FZS, a few highlights include Vietnam's government, the San Diego Zoo, Australia's Melbourne Zoo, and Germany's Allwetterzoo Münster, among others. The EPRC's guest book includes luminaries from every wildlife organization I can think of, and among the handwritten notes I see one that seems to practically beg Nadler to ask them for money.

Back in '98, there was one animal keeper besides Tilo and Hien. Now there are twenty animal keepers recruited from the local villages, a veterinarian, a biologist, an operating room, and an infant-care center for the primates, in addition to the old veterinary clinic and the old quarantine center. Whereas Hien used to conduct English-language classes for the animal keepers by herself, there are two volunteers taking over that duty. There's also a regular roster of graduate students doing studies here to gain experience for their zoology degrees.

Scientists have come here to study the EPRC's primates from facilities such as the University of Colorado at Boulder, the Kansas City University of Medicine and Bioscience, Johns Hopkins University, and Ohio University's College of Osteopathic Medicine. They seek answers to questions such as how each type of primate moves or "locomotes"—for example, do langurs walk on their knuckles as apes do, walk like us humans, or swing through the trees like gibbons? (To watch gibbons swing from branch to branch is a wondrous thing: they swing, pivot, leap, twist . . . and then stop abruptly, before performing more aerobatics. Some suspect that they have a deceptively specialized anatomy—including flexible shoulder joints and long, hooked fingers—that enable these behaviors.)

Other questions have also been tackled. In cooperation with the German Primate Center, the taxonomy of Vietnamese primates has been clarified on a molecular genetic basis, which is an important step for their successful conservation, as such information is vital to any breeding program. New species such as the gray-shanked douc langur (*Pygathrix nemaeus cinerea*) and the Annamese silvered langur

(*Trachypithecus margarita*) have been identified. Food and nutrient studies have been carried out with biologists from Hanoi National University; the University of Forestry Xuan Mai, Vietnam; the University of Cologne, Germany; and George Washington University in the United States.

Tilo Nadler has had a hand in innumerable publications, surveys, and studies of the animals' natural history; and the primatology community's collective knowledge about the needs of langurs alone has increased, covering everything from food requirements to environmental and health conditions.

The physical conditions they work under have changed too: workers at the EPRC no longer have to take a bus to the capital city to e-mail a report or send a fax but now can use any of a number of computers hooked up to the Internet here in the hinterlands. Nadler's office alone has three computers and a voltage stabilizer, along with the requisite phone lines, printers, DVD burners, and memory sticks. In case the power goes out, there's a diesel backup emergency generator on the premises. The EPRC also has its own website, www.primatecenter.org, which contains updated listings of the latest volunteer opportunities, and an "Adopt-a-Monkey" [*sic*] program. And, of course, the Nadlers themselves now have a house in which to live, as opposed to renting a single room outside the park boundary like they first did.

The park has also evolved. In addition to regular bus service, Cuc Phuong National Park has a museum collection, an orchid nursery, an ecotourism village, color brochures, bicycles for rent, and guided night tours.

Probably the biggest change is that the EPRC has spun off some direct ancillary rescue projects. These include a Turtle Rescue Center and a center devoted to saving civets and pangolins (a catlike night hunter and a scaly anteater-like creature, respectively). There are also longer-term release sites such as Van Long Nature Reserve and Phong Nha-Ke Bang National Park.

Nationwide, there are now many wildlife rescue projects, scattered up and down the country; these include the efforts of Bill Robichaud to save the saola; that of Le Khac Quyet to save the Tonkin snub-nosed monkey, Alexander Monastyrskii's work to save butterflies, Leonid Averyanov's efforts on behalf of rare orchids, and Jeb Barzen and Tran Triet's project to restore the Plain of Reeds and its cranes—to name but a few. A surprisingly large number of these projects bear some relation to the EPRC. The variety of wildlife initiatives is extensive, and includes the Bear Rescue Center in Tam Dao, a sea-turtle rescue center on what used to be Vietnam's version of Devil's Island at Con Dao, and numerous other centers where researchers are trying to protect beetles, paddlefish, orchids, and all manner of other wonderful species against great odds.

"On the plus side, there are more Vietnamese organizations and more international organizations interested than before," Tilo says. "Fifteen years ago, there was really no tourism here. Now there's been an explosion of tourists to Cuc Phuong. Unfortunately, while there is more interest, there is not more real understanding. Tourism in Vietnam means entertainment—a loud party in the woods. I fully understand the need for such an outlet, but it should not be in a national park. . . . It is not the duty of a national park to provide everything for everybody all the time; some places need higher levels of protection. This is not an amusement park."

Hien agrees, saying, "They ask us, 'Where is your wildlife restaurant, we want jungle food.'" She makes a face and adds, "You see women showing up here in high heels, and men in business suits, expecting to go into the wilderness."

Tilo adds that this attitude of parks as playgrounds is encouraged by the government, which sees the parks' tourists as a lucrative source of hard currency. Officials want to make Cuc Phuong National Park part of a grand tour for city people wanting to see the sights in the region. So, the government has built things in the park that really have nothing to do with conservation or preservation, such as a fifty-meter swimming pool that has never been used—"It's completely green, completely useless, and there's often dead animals floating in it," Tilo says. There are also huge, empty museums, unused badminton courts, and an expensive training facility that is only used twice a year—one of three such training facilities here.

There are also things that are overbuilt: there were six large conference centers built inside the park recently, each of which can hold more than a hundred people at a time—far more than ever needed.

From what I've learned, such overbuilding is common in developing countries; few things symbolize Progress with a capital "P" more than a building or a road or a dam. More important, these hard-to-miss, impressive physical structures give politicians something they can point to as accomplishments. The same is true for many of the Western NGOs that fund some of these projects; they want pictures of something to publish in their annual reports. In contrast, it's hard for constituents to tangibly see that there is more biodiversity in the landscape.

Nadler expounds on the theme, saying, "Maintenance is the biggest problem here. It is much easier to write a proposal for a new building than for paint." He laughs, and adds, "The little everyday things are the hardest here. We need thousands of screws, and we just can't get them."

From a wildlife-rescue point of view, the Nadlers said, money is "better spent on conservation and preservation of the land the park already has"—for items such as more rangers, more habitat protection, and more law enforcement. From their experience with the park, however, most money is spent with an eye toward quick-and-dirty projects that draw in more tourists.

Probably the most egregious case concerned the handling of a small group of minority ethnic Muong people who had lived for generations within the park boundaries, at a place called Mac Lake.

Located in the middle of the national park, their village consisted of about five traditional family houses, all of which consumed a small amount of firewood and other park materials. The people were evicted from Cuc Phuong under the guise of stopping firewood harvesting; several expensive tourist cottages were then erected in the remains of the old village. The money went to fatten the park's coffers, while the indigenous people gained nothing. And the fire-wood consumption at the tourist facility is ten times higher than it was when the Muong lived there, declared Nadler.

To increase funds, bureaucrats even require volunteers to stay in overpriced housing within the park and pay for it out of their own pockets; the attitude rings of what an Australian friend had first warned me: "The semi-official slogan of Vietnam is 'Charge the foreigner triple.'"

As for the status of the wildlife nationwide, aside from langurs, Tilo says that results have been mixed. The rhinos at Cat Tien National Park are extinct. There were twenty-four tons of endangered pangolins (*Manis pentadactyla*) seized from smugglers in a single shipment in Haiphong; the animals are a delicacy, and the wild ones bring in the most profit. "Bushmeat is always cheap; all you need to pay for is the cost of one bullet," says Tilo. Where there used to be hundreds of tigers in Vietnam, he now estimates that there are only thirty—although when dealing with elusive animals that are hard to census, researchers are prone to disagree on exact numbers, other than that they are very low.

Asked about the number of elephants countrywide, Tilo says that there are now about 120 in the wild, all in widely scattered groups of two or three—not enough to be a real herd in any one place. Historically, Vietnam was probably once home to large numbers of the creatures, as can be seen by old documents and eighth-century temple facades, which are festooned with carvings of Asian elephants. Tilo and Hien disagree as to the size of the original population; Hien guesses that there were originally ten times the number that there are now, while Tilo guesses that there were originally one hundred times more. In any case, that means that there were thousands or tens of thousands of elephants in Vietnam.

"Are we making progress?" someone asks.

"I'm an optimist, but only if we have real government support to protect our special places," Tilo said. "Sometimes you need to make tough decisions; this may mean sacrificing a park outside Sa Pa [a mountain resort town close to the Chinese border] to a dam, in order to save something somewhere else." He rattles off the pluses of the past two decades: better law enforcement, slightly better regulations, and more education projects. Against that, he lists the negatives: a fast decrease in the numbers of wildlife, and increased habitat loss. (His attitude is echoed by outside zoologists in other parts of the world, such as Alan Rabinowitz, who said, "I see setbacks constantly, but at least I can look around me and know that vast, beautiful reserves are out there that would not have been there otherwise. . . .")

Even when habitat is not completely wiped out, it can still be so sliced and diced by roads as to be barely usable by wildlife. Running a busy superhighway through a park can mean cutting off migration routes, or sending a thick stream of overloaded trucks and zooming buses through a herd. Rare animals get run down, poachers get in and out more easily, and invasive species hitch rides and spread throughout the park.

Tilo cites the case of a pair of Vietnamese reserves located a short distance apart. One of them, the Kon Chu Rang Forest Complex, has a large number of rare and endangered tree species. Meanwhile, the other, Kon Ka Kinh National Park, has a large population of endangered gray-shanked douc langurs. These animals feed upon the very trees found in Kon Chu Rang, so it would make sense for the two reserves to be joined together into one single, large, protected block. Both would benefit: the langurs would have a larger food supply, and the forest would have something to help spread around its seeds and fruits. Instead, the government plans to build three dams, a bauxite mine, and a new road between them, thus forever dividing the two sites.

At the same time, he can understand the needs of the local people, who want the latest goods; faster travel to the hospitals, markets, and services of the big city; and a better shot at getting the basics that we Westerners take for granted, such as electricity or mobile phone service. They also want jobs, and a piece of the economic pie.

Nadler understands this need but says that there is a time and a place for everything, and that certain places, such as national parks, are best suited for one thing only. "A park cannot provide all the resources at the same time," he says.

Efforts to protect the wildlife and provide a living for the local people have to be very carefully considered, he said, citing the Mac Lake example. Even the operation of the "buffer zone" between the border of a park and its nearest city has to be examined; people often think of this area as a good halfway point to conduct "mixed use" of natural resources—that is, a region not as sacrosanct and rigidly protected as the park itself, but not a fully urbanized and developed area either. Around here, there's about fifty thousand people living in the buffer; they collect snails, mushrooms, firewood, animals, and birds, and graze their livestock.

The temptation has been to provide more economic opportunities in such buffer zones, in an attempt to provide the people in them with a living wage. The problem, says Tilo, is that "if you make things too good in the buffer zone, then people don't want to leave, and it actually attracts more people and puts more pressure on the park." For example, at Pu Mat Nature Reserve near Laos, there were 250 people per day cutting down trees in the reserve for firewood. After a €20 million project by the German Development Bank to aid people in the Pu Mat buffer zone, there are now 700 people per day harvesting firewood in the reserve. "The development efforts must be far enough away geographically to pull people *out* of the parks and out of the buffer zone. If it is right in the buffer zone, then people move into it." Nadler sighs. "No one studies the success or failure of these projects."

Which makes it all the more astonishing to think of all the things the EPRC did accomplish. How did they do it? Over the past several days, Tilo has constantly reiterated the things that he thinks are key: clearly defined, highly targeted, and specific goals; coordinating the saving of animals with the saving of their habitat; building a rapport with the local people and giving them an economic stake in the success of the creatures through such things as ecotourism or the direct hiring of villagers for rescue programs. He says it is important to make sure that one program—such as bringing opportunity to local villagers—does not interfere with other goals, such as saving the integrity of the park as a single entity. For creatures that are on the very cusp of extinction, he emphasizes the use of the stick as well as the carrot, and has no qualms about being tough. Tilo also understands the rough and tumble of politics, and that sometimes one nature reserve has to be sacrificed so that another may survive.

It's also apparent that any serious effort takes a large investment of time, in his case the better part of two decades. (And it helped that because of the accident of his nationality, he had a several-year head start.) Growing up as he did under the old East German Communist system, Nadler learned early on how to manipulate the Party to get what he wanted in spite of officialdom—a handy skill indeed in the Socialist Republic of Viet Nam. Tilo Nadler was supposed to be a refrigeration engineer, but somehow managed to get a degree in biology, work on conservation projects, and take part in scientific expeditions worldwide before coming to Vietnam.

And Nadler's informal, casual, and sometimes earthy style—if sometimes abrupt and direct—overall meshed well with the Vietnamese, whose culture stresses informality and humility. Some famous ancient Vietnamese temples were purposely constructed of fragments of recycled and broken construction debris, to remind the faithful of the importance of being humble.

Before the evening ends, however, it's apparent that Tilo also had a secret weapon that contributed to the EPRC's success: a member of the "long-haired brigade."

Looking deceptively slight and fragile in their *ao dai* and long flowing black hair, and appearing shy and delicate on first sight, Vietnamese women are nevertheless very tough survivors, skilled at organization and negotiation. During the twentieth century's wars in Indochina, they played vital roles in ferrying supplies, running clinics, aiding in hospitals, and were even included in construction battalions. Referred to as "the long-haired brigade," the name stuck.

Ho Chi Minh himself supposedly credited Vietnam's women in the country's push for independence; women did much of the work of running the revolution behind the scenes. In doing so, they were keeping to an ancient tradition.

Women in Vietnam have long had the same legal rights as men and could own property, dating back to a time when their sisters in Europe had no such rights at all. Young Vietnamese girls also had strong female role models; some of the country's earliest nationalist heroes were women, such as the Trung sisters, who led an army against foreign invaders. Or Trieu Au, who was a sort of third-century Joan of Arc figure, leading the fight for independence from the Chinese. Legend says she declared, "I wish to ride the tempest, tame the waves, and kill the sharks." All these women warriors are idolized, with national holidays and temples in their honor. They are often depicted on statues as riding war elephants into battle.

The description may sound like dated socialist-realist propaganda, but the women to this day reflect these characteristics. If a modern traveler ever has to bargain with one of the Vietnamese women for the price of a meal or a hotel room, then they must be prepared for a tough battle. With the men, there is generally more give-and-take, but with the women, you might as well throw in the towel and surrender.

Tilo's own "long-haired brigade" member is Hien, who fits the description despite having her hair cut in a more modern, short hairstyle and wearing Western fashions. This daughter of a Communist Party official from outside Hanoi is a whirlwind of activity: fluent in three languages, she organizes, administrates, and handles all the logistics and office politics—not to mention doing the cooking and raising the children—thus freeing Tilo up for doing the lengthy biological surveys and rescue fieldwork.

She was the one who immediately took over all my plans for my stay in the park, and arranged (or commanded) my meeting with certain Party officials. She also coached me on what military officer I should ingratiate myself with, and how to play them off against each other. My diary says that "Hien can be quite bossy and officious at times, but it is apparent that she does it all for the benefit of Tilo and the EPRC . . . I think he likes her fussing over him . . . married life and fatherhood seem to agree with him, and he smiles a lot more than the last time I saw him."

With dinner over and the picture taking done, the subject of how they became a couple came up. Hien tells us that she had more or less fallen into this by accident; with her bent for the German language and her valuable Party connections, she was a natural to be Tilo's interpreter. Because they were spending a lot of time together, they got to know each other well, and Hien discovered that the more she knew about Vietnam's wildlife, the more she wanted to protect it as well. It must have been a tremendous ideological jump, but she put all her skills to use in helping the EPRC get off the ground.

Hien told us that it was more of a battle to convince her parents of Tilo's suitability as a husband; she said that she had to wait until her father was retired and his career in the Party was no longer at risk before she actually tied the knot with a foreigner.

At the same time, Hien describes herself as a very atypical Vietnamese girl growing up; she had a more independent streak, and enjoyed things like splashing in mud puddles and climbing trees. She fumbles for the right expression in English: "A tomboy?" I ask. She does not know the phrase but nods yes after I explain.

When asked about how she happened to speak German, Hien is more reticent. But in rereading an interview from a dozen years ago, I remembered what she said about how she picked up the language. As an infant, she was injured when downtown Hanoi was bombed by US Air Force B-52s during Richard Nixon's "Christmas Bombing" of 1972, also known as "Operation Linebacker II."

The raids, which took place from December 18 to 29 of that year, have been called the largest heavy bomber strike launched by the US Air Force since the end of World War II; they were conducted as part of a strategy to force Hanoi to the negotiating table and thus end the Vietnam War (or what the Vietnamese called the American War) on the best possible terms for America's South Vietnamese allies. About forty thousand tons of bombs were dropped "in the most concentrated air offensive of the war against North Vietnam—and the episode still arouses controversy," wrote Stanley Karnow in his book *Vietnam: A History*. The targets were primarily railroads, power stations, docks, shipyards, radio transmitters, and similar facilities; the intent was to send a signal of toughness to America's adversaries.

Nixon told his administration that he was loosening up any restrictions about what was or was not a legitimate military target. He told the Joint Chiefs of Staff: "I don't want any more of this crap about the fact that we couldn't hit this target or that one. This is your chance to use military power to win this war, and if you don't, I'll hold you responsible."

About 1,600 civilians died in Hanoi and Haiphong in the raids.

The casualties were not on the scale of Hiroshima, in which hundreds of thousands died; and most of Hanoi's buildings were spared, as I saw for myself decades later. As Karnow—who was a reporter stationed in Vietnam at the time—stated, "In fact, the B-52s were programmed to spare civilians, and they pinpointed their targets with extraordinary precision. Nevertheless, some bombs did stray, with ghastly results."

One of the bomb victims was Hien, who was immediately flown to East Germany, a Communist-bloc ally of Hanoi, for medical attention. A child at the time, she used her long recuperation there to learn how to speak German, little knowing that years later her language skills would allow her to become the interpreter for a German man who wanted to rescue Vietnam's endangered species.

Remembering the story and the view of the old craters from the plane, it raised the question of how all those animals in the wild survived all that destruction.

11

Surviving "American War": Agent Orange

There was a special Air Force outfit that flew defoliation missions.
They were called the Ranch Hands, and their motto was:
"Only we can prevent forests."

—Michael Herr, *Dispatches,* describing the siege of Khe
Sanh just before the Tet Offensive, Vietnam, 1968

WHEN YOU VISIT Khe Sanh—the former battlefield where US Marines held out in 1968 against a four-month-long siege conducted by massive numbers of North Vietnamese troops just before the Tet Offensive—there is surprisingly little to see. It is now mostly a coffee plantation, along with a small museum and a guestbook. Off to one side is a single rusting 105mm howitzer that serves as a monument. Otherwise, every piece of metal has been torn up, dismantled, and sold to scrap-metal merchants; sometimes, it seems as if the locals consciously eradicated any evidence of the American or French era.

In the poignant words of Michael Kelley, the author of *Where We Were in Vietnam:*

> Returning U.S. veterans now often find themselves both astonished and befuddled by the almost total absence of our former presence. Places they thought would be simple to locate are instead almost impossible to position under the heavy footprints of intentional erasure, population growth, cultivation and the inexorable process of Mother Nature as artist and sculptor. "There" is simply no longer there, in many instances.

Despite this, I had wanted to see for myself what remained of the old dividing line between North and South Vietnam, and learn what influence such battle zones had on today's environment and its wildlife in Vietnam. So I had gone on a bus tour conducted by the local tourism agency out of Hue.

Such tours are popular with Western visitors. Our guide is an English-speaker, Ngo Dzun, who was born and raised in the village of nearby Dong Ha. He tells us that during America's war in Vietnam, he was too young to serve, although his brother was drafted—which is how Ngo learned the language.

As our tour bus of about a dozen people rolled along from our meeting place to Quang Tri and the real start of the tour, Ngo informs us that the road we are on—Highway 1—had once been infamous for ambushes, earning it the nickname "Street Without Joy."

After a brief stop, we turn west on Highway 9 and head into the hills toward Khe Sanh. Our tour takes us past such legendary battle sites as "The Rockpile." From this point to just beyond Khe Sanh, the US Air Force dropped about 100,000 tons of bombs

in seventy-seven days in support of the Marines who were surrounded there.

They also dropped huge amounts of Agent Orange to strip the
land of its cover. We tourists see lots of stunted vegetation around
the Rockpile and other areas; in general, the farther away we are
from the battlefield, the more trees we see and the taller the vegetation. Some of these trees are acacias, a fast-growing, non-native
genus from Australia, planted by returning US and Australian vets
in an effort to help restore the land of Vietnam. There are many
such efforts from former enemies; Ngo tells us that an American
veterans association came here to train the locals how to safely remove mines. Some even brought maps of old minefields.

When we get past the Rockpile and up into the hills, all we can
see are bushes, vines, and grass; there are no trees. There is, however, a lot of mud and rain—this is near the notoriously wet region
around the country's middle. A man in a black poncho herds water
buffalo in front of our bus, and we have to wait for the animals to
pass before continuing. At the battlefield of Khe Sanh itself, there is
not much to see except for a barren, lifeless longitudinal stretch of
red dirt that they tell us was the old landing strip; the whole site is
surprisingly disappointing due to this lack of visual remains. There
is a strong smell of wood smoke in the air from the nearest human
settlement, mixed in with the overcast. After touring the old battlefield for an hour or so in pouring rain, we climb back aboard, visit a
local Bru tribal village, and make a quick side trip to the boundary
with Laos to drop off some passengers. We then head back to the
area where Ngo grew up.

It is easy to be fooled by what we see as we head back toward
the coast. I notice a mountain that shows scars, then am told that
these are a recent occurrence, caused by the blasting of a rock and
gravel excavation company. The scenery in general looks like the
low scrubby, treeless California hills of the old *M*A*S*H* comedy

series. Is this natural? Caused by napalm? Agent Orange? Traditional slash-and-burn agriculture?

Ngo says that many families in the region were afflicted with birth defects, in the area past the Rockpile and about ten kilometers west of Dong Ha. He cites many cases of spina bifida, saying that they were related to the consumption of Agent Orange. Somewhat unexpectedly for someone from the DMZ, Ngo says, "I think that Agent Orange is no longer in the water now, twenty-five years later," emphasizing the words "water" and "now." "But there may be dioxin left in the land."

His concern is that immediately after the war, there was a lot of Agent Orange in the environment—for which the people were not treated—and he is personally convinced that there are residual effects. "They say that Agent Orange didn't affect many people. That's not true," claims Ngo. "There were many problems. Many got sick or had abnormal babies. Even today, nearly two hundred abnormal babies are born in the area. Even now, local people get sick because of what happened here." As for its effect on the land, the fish, the birds, and the domestic and wild animals, he cannot give a definitive, absolute answer, and neither, apparently, can anyone else. Ngo simply points to the landscape and says, "You see no vegetation."

Back in Hanoi, the country's most renowned biologist, Vo Quy, is more cautious about making any flat declarative statements about the effect of Agent Orange on humans, saying that it is hard years later to pin down any visible long-term, repeated, continuing insults caused by defoliants like Agent Orange. But he is personally convinced that the chemical's effects are ongoing when it comes to the landscape, calling it "ecocide" and saying that many large areas affected by herbicides are still unsuitable for cultivation or livestock.

In testimony to the US House Foreign Affairs Committee in 2009, he said, "War does not end when the bombs have stopped falling."

To get a sense of Agent Orange's impact on the wildlife and environment, you have to try to extrapolate from what little knowledge we have about the chemical, nearly all of which was laboriously, carefully, and slowly extracted from retrospective studies designed to examine the effects of different amounts of Agent Orange upon humans. Much of that information is still contentious politically.

The US Air Force dropped about twelve million gallons of Agent Orange alone between 1961 and 1971, in an effort to deny cover and food to the enemy. In addition to this well-known herbicide, the Air Force also dropped lesser-known chemicals such as Agents Purple and Pink; all three contained the contaminant known to chemists as 2,3,7,8-Tetrachlorodibenzo-p-dioxin, popularly called "dioxin." The compound was a by-product of the manufacturing process, and it was thought harmless at the time, although it is increasingly understood to be toxic to humans. The World Health Organization (WHO) says: "Once dioxins have entered the body, they endure a long time because of their chemical stability and their ability to be absorbed by fat tissue, where they are then stored in the body . . . [dioxin's] half-life in the body is estimated to be seven to eleven years. In the environment, dioxins tend to accumulate in the food chain. The higher in the animal food chain one goes, the higher the concentration of dioxins." What the WHO said applies to all mammals at the top of the food chain, including humans, of course, but also all other creatures.

In addition to these dioxin-based compounds, the US military sprayed other compounds as well, including Agent Blue, a form of organic arsenic, targeted at destroying enemy crops such as rice. They also spread herbicides that did not contain dioxin, known as Agents White and Green, after the color of the barrels they came in. Altogether, the different substances make for about twenty million gallons of chemicals sprayed over portions of Vietnam.

The long-term effects of this rainbow of defoliants on the country's wildlife are still largely a mystery, with most of the concern centered on those defoliants that contained dioxin. We have the most information about dioxin's effects on humans. We know much less about its effects on domestic animals. We know next to nothing about their effects on Vietnam's rare wildlife.

The reason? There is a tremendous financial pressure on researchers to focus on domestic animals or commercially valuable wild creatures first. For example, for every study of a rare, wild species like, say, the whale shark, there are probably dozens, if not hundreds, of research studies on commercially important fish such as cod, lamented shark researcher Dennyse Newbound of the University of Western Australia at Perth. Pure research for the sake of research is often at a disadvantage.

Essentially, there is a tendency for lawmakers, grant-givers, and research institutions to be most concerned about species we consume every day, and less interested in general in those species that are more tangential to our daily lives, a tendency that is reflected in funding priorities. But given that researchers combing through old archives found that about four times as much dioxin was sprayed in Vietnam as previously thought, the prognosis is not the best.

As of June 2011, after much delay and acrimony, Vietnam and the United States have finally taken the first steps toward a joint effort in cleaning up dioxin contamination. For years, the two countries had argued over the problem, because it could lead to claims for financial compensation; the BBC described the situation as "the biggest remaining war-era issue." In a nutshell, the US government position was to deny that the chemical had any lingering effects at all, while the Vietnamese held the opposing view. And tied in to it was the looming cloud of financial compensation for damages caused by the war. Over the passage of time, the US position has evolved, and the Department of Veterans Affairs website now says:

"Studies suggest that this chemical may be related to a number of cancers and other health effects in humans."

Scientists have found dioxin remains in freshwater animals decades after the war. The most likely common scenario is that the chemical collects at the bottom of the country's many farm ponds; there are about fifteen million bomb craters in Vietnam, many of which filled with water and are used for aquaculture. Most likely, when fish or ducks stir up the mud at the bottom of the ponds, they ingest the dioxin and it gets into their fatty tissues, where, due to a quirk of its chemical structure, it stays more or less locked in place without breaking down, in a manner similar to DDT. As you climb up the food chain, the level of dioxin is magnified greatly, for the simple reason that if one tadpole contains one unit of dioxin, and one small fish eats ten tadpoles, then that small fish will now have ten units of dioxin. A human eating ten of those contaminated fish would ingest one hundred units of dioxin, just from that one meal.

Such "bioaccumulation" or "biomagnification" is a well-known phenomenon, well recognized in science and described in environmental textbooks. For years, the question was not whether the level of dioxin built up in this manner in higher mammals but whether dioxin was linked to negative effects, such as a statistically significant increase in the number of cancers, birth defects, and other ailments among those with high levels of the chemical in their bodies.

Few studies have been done with specific animals in mind; most have tried to look at the big picture in the overall environment, with an emphasis on those at the top of the ladder: we humans.

The issue became a contentious one, much like an earlier era's battle over whether cigarette smoking causes lung cancer. Though widespread spraying of Agent Orange (and the dioxin contaminant it contained) began in the early 1960s, it was not until 1994 that the US Environmental Protection Agency labeled dioxin as a possible carcinogen. Over time, more and more studies showed that

Above Cat Ba golden-headed langur. Depending on who you talk to, there are about sixty or eighty of them left. It was only discovered and named about ten years ago, in the center of Ha Long Bay, northeast of Hanoi. PHOTO BY TILO NADLER, ENDANGERED PRIMATE RESCUE CENTER (EPRC)

Right Tilo Nadler, founder of the EPRC, photographed in his office. PHOTO BY DAN DROLLETTE

Above The best way to spot Delacour's langurs in the twelve-square-mile nature reserve of Van Long is by boat. The animals can usually be found on the sides of the steep islands jutting up from the water. About a hundred of them are known to be in the wild, with most living at Van Long. PHOTO BY DAN DROLLETTE

Right A civet, a type of wild cat that hunts at night, in a facility for small carnivores at Cuc Phuong National Park, which was patterned after the successful EPRC. The animal must take the prize for most unusual commercial uses: after raw coffee beans pass through its digestive system, the result is gathered up and processed into the most expensive gourmet coffee in the world, at a reported $300 per pound. Similarly, the animal has an anal gland that imparts a scent highly desired by industry; the gland has long been the basis for many expensive perfumes. PHOTO BY DAN DROLLETTE

Below Vietnam's endangered Eatern Sarus cranes, photographed by a Vietnamese wildlife biologist on the Plain of Reeds in the Mekong Delta, where rescue work brought these animals back from near-extinction. PHOTO BY NGUYEN VAN HUNG, TRAM CHIM NATIONAL PARK, DON THAP PROVINCE, VIETNAM

Center The kouprey, one of the region's most eagerly sought animals, was discovered by accident by a zoologist at a Paris zoo who was examining a shipment of wildlife from Indochina in 1937. Unfortunately, the world's only captive specimen of a kouprey died during the German occupation of Paris in World War II. This is one of the only images anywhere of the animal, and the only known one of it in captivity.

Bottom of page The Annam flying frog is reputed to get around the forest by spreading its webbed toes and leaping from tree to tree like a flying squirrel. PHOTO COURTESY OF NGUYEN HAO QUANG, CENTER FOR BIO-DIVERSITY AND DEVELOPMENT, HO CHI MINH CITY, VIETNAM

Opposite National Tropical Botanical Garden's "extrem[
botanists" Ken Wood (foreground) and Steve Perlm[
(background) rappel down cliffs above the Kalalau Va[
ley in Kauai, Hawaii, in order to save highly endanger[
tropical plant species from extinction. Wood is holdi[
a flowering plant between his teeth. PHOTO COURTESY [
CHRIS JOHNS, *NATIONAL GEOGRAPHIC*

Left The rare *saola,* an animal discovered only in t[
early 1990s, setting off a "biological gold rush." It [
known to local tribespeople as the "polite animal" for [
dainty, quiet way of slipping through the forest. ARTWO[
COURTESY OF PATRICIA J. WYNNE

Below A Giant Mekong catfish, one of many large fi[
that have been found in Indochina by Zeb Hogan, a N[
tional Geographic Explorer in Residence who specializ[
in seeking out new and rare fish species. Critically e[
dangered, this catfish migrates up the Mekong River [
Laos and northern Thailand to spawn. PHOTO COURTE[
OF ZEB HOGAN

Above An Asian black bear being illegally "milked" for its bile in a painful process at an underground lab [
Hanoi. This photo is part of a prize-winning series on the black market trade in animal products, photograph[
by Mark Leong, who won the Wildlife Photographer of the Year award. The bile is marketed as a health cu[
although there is little scientific evidence for this claim. The demand for animal products has skyrocketed w[
increased wealth—much of the appeal coming from the fact that using such things is seen as a status symbol amo[
the newly rich. PHOTO COURTESY OF MARK LEONG/REDUX PICTURES

Right This five-foot diameter softshell turtle, a specimen of *Rafetus swinhoei,* is probably one of Vietnam's most endangered species, and one of only four individuals of this species in the world. It dwells in the center of a lake in the middle of downtown Hanoi. PHOTO COURTESY OF NA SON NGUYEN, ASSOCIATED PRESS

Above Nothing like getting your feet wet! Jeb Barzen (center), self-described as "just a duck biologist," working to restore bird habitat on the Plain of Reeds in the Mekong Delta, in the south of Vietnam. On the staff of the International Crane Foundation (ICF), this American biologist has been collaborating with his Vietnamese colleagues for decades. PHOTO BY TRAN TRIET, ICF

A father, mother, and baby Delacour's langur. Found only in the limestone-rich mountains of Indochina, the babies always start out bright orange, turning black as they grow up. PHOTO BY TILO NADLER, EPRC

the chemical was much more dangerous than first thought; in the beginning, it was considered completely harmless, and there are stories (possibly apocryphal) that in the early days of the Vietnam War, soldiers sometimes drank Agent Orange–laced water in order to prove how "harmless" it was.

Similarly, an American soldier serving in Korea in the late 1960s, Phil Steward, claimed, "We were told, 'You can drink it, you can brush your teeth with it, or you can bathe in it. It won't hurt you.'"

Then the pendulum swung the other way, and it sometimes seemed that exposure to the contaminant could cause anyone to be struck down, including the son of the same US admiral who ordered the spraying of Agent Orange in Vietnam in the first place. Elmo R. Zumwalt III was exposed to the chemical while serving in Vietnam, later coming down with cancer and dying in 1988. Reflecting upon the difficulties of showing cause and effect, in a 1986 article for the *New York Times Magazine,* Zumwalt said: "I am a lawyer and I don't think I could prove in court, by the weight of the existing scientific evidence, that Agent Orange is the cause of all the medical problems—nervous disorders, cancer and skin problems—reported by Vietnam veterans, or of their children's severe birth defects. But I am convinced that it is."

With all this history, one would think it would be easy to at least see where Agent Orange was physically sprayed in Vietnam. But it is hard to tell just by looking at the landscape. All a casual visitor sees is . . . a group of ducks in a farm pond, or grass on a hill where locals say there once used to be trees.

The best way to get a handle on it all seemed to be to talk to a woman who is probably one of the best sources of detailed information on the spraying of Agent Orange, and whose work was cited by Vo Quy, veterans' groups, and the Ford Foundation. She is an academic who works not in the jungles of Vietnam but among the towers of Manhattan, almost within the shadow of the George Washington Bridge.

. . .

Dr. Jeanne Mager Stellman is an epidemiologist at Columbia University's Mailman School of Public Health, who has won numerous awards and fellowships for her work, including a Guggenheim. Despite the numerous plaques on her office wall, Stellman was remarkably low-key and informal. Under contract to the National Academy of Sciences' Institute of Medicine, she's been leading a team that is trying to evaluate herbicide exposure on those who were in Vietnam.

After I interviewed her, her team was to complete the first detailed, computerized map of the patterns of herbicide spraying in Vietnam with precise target zones. Titled "The Extent and Patterns of Usage of Agent Orange and Other Herbicides in Vietnam," their work made the cover of the peer-reviewed scientific journal *Nature* in 2003—a blue-ribbon mark of honor in science from one of the top two journals, an event that is the scientific equivalent of a twenty-one-gun salute with a ticker-tape parade down Broadway. Users of the map can now know how much defoliant was sprayed where and when on whom. And, by extension, users could make an educated guess as to what was done to the environment and its wild denizens.

In the course of doing research to create the map, the team also found that the amount of dioxin sprayed—the crucial contaminant—was up to four times greater than previously estimated.

The completed job sounds simple and straightforward, leading some to ask why this was not ever done before. But as with many "obvious" things, the devil was in the details: the work takes into account daily logs filed by pilots after missions, flight-path information, the amount and type of agent dropped, troop locations and movements, known locations of the Vietnamese population, land features, and soil types. It also allows for whether the defoliant was deliberately released in a gradual, evenly spread, controlled manner or dropped in short giant bursts as the result of aborted missions, crash landings, emergency dumps, leaks, midair collisions, or accidents. The study goes so far as to include a section on the cleaning and disposal of discarded drums of herbicide, which typically still held about two liters of liquid in the bottom of the drum after each one was "emptied." (Given what I saw at modern-day, postwar Khe Sanh, where every scrap of metal was scavenged for sale, empty drums of herbicide would have been avidly recycled for other purposes, with who-knows-what effects.)

At the time we met, Stellman was still in the midst of the detective work. She explained, "The basis of epidemiology is person, place, and time. That tells you how much they were exposed to, and then you can go from there to see if there is a link between herbicide spraying and the health of American veterans and the Vietnamese." Her goal was to go beyond anecdotal evidence, and put numbers to things to see if there was indeed a cause-and-effect relationship.

Cause and effect can easily be confused by observers; as the saying goes, "Just because the rooster crows in the morning doesn't mean it makes the sun rise." Or, if you like: "Association is not causation."

Vietnam has a number of what are called Peace Villages—I visited one after the tour with Ngo Dzun—in which people with all kinds of illnesses and diseases were lumped together. The administrators of the Peace Villages liked to say that everything you saw was caused by Agent Orange, but critics—rightly or wrongly—charge that the Peace Villages had become a catchall for any type of condition; this was a point reiterated by Ambassador Peterson. The Peace Villages did not constitute evidence all by themselves for the malevolent effects of dioxin; for example, they did not take into account the effects of chance, malnourishment, or the conditions of prolonged warfare.

Rather, what scientists want is more demanding and more concrete. "We may know that a plane flew over a given plot of land at a given time," said Stellman. "But what we want to know is what the 'exposure opportunity' was—the first step to making firm links with the health data." The goal was to produce the user-friendly database that tells users "This is what was sprayed at that time, and this is where you were at that time." The user can then compare this with what subsequently developed in terms of their own health.

Previous studies did not ask the right questions. Most studies just inquired about the health of *all* the soldiers after returning from Vietnam, without regard to where they had been located or what they were doing—sitting in an office doing clerical work far from the front lines, or getting doused with herbicides in the jungle. (Like in all wars, there were many more support staff than combatants. Four out of five men in the US Army in Vietnam did not go near combat, Stellman said. Similarly, in World War II, "fighting men comprised ten percent, or less, of the full military complement," wrote military historian Gerald F. Linderman in *The World Within War.*) Stellman maintains that what the studies should have asked was whether combat soldiers exposed to high levels of dioxin developed high levels of health problems later.

To attempt to do all this was enormous work. Just finding out who was over there was a big detective job: a lot of material was incomplete, and she said that her team had to pull together the first comprehensive list of military units sent to Vietnam. No such list existed previously; there were a lot of covert actions, many different military units, and no pressing need for a grand, overall enumeration at the time. Reconstructing it after the war's end made for a horrific job.

Researchers were fortunate to have any data at all. Many records were nearly lost when the US embassy in Saigon fell; luckily, an officer got the reels of computer tape out on one of the last helicopter flights. Even if data could be found, military units moved around frequently, which may or may not be reflected in the written record. Other units may have been understrength or over allotment. And there may have been unrecorded defoliation as well. Stellman said that the more she delved into it, the more the team of scientists could appreciate the complexity of the logistics of running a war.

Despite the breakthroughs, she felt that there would always be gaps in our knowledge about precisely what was dropped under battle conditions in the forests of Vietnam. The overall picture is getting clearer, but the minute, specific details of a particular place will always be subject to doubt, much like a weather forecaster can know what will happen region-wide when a storm is about to hit, but cannot say with certainty what will happen in your immediate neighborhood in the next hour. It will probably always be a dicey question as to precisely how much dioxin fell on a particular part of the country, which means that we will never be able to be 100 percent certain of its effects on humans—let alone on wildlife, plants, or the environment. And what we do know about its effect on people is still being deciphered, although a consensus seems to be forming.

Probably the best summary comes from the World Health Organization, which says that dioxin—the worrisome chemical contaminant in Agent Orange, and a by-product of the manufacturing process—is "a persistent environmental pollutant that is highly toxic and can cause reproductive and developmental problems, damage the immune system, interfere with hormones and also cause cancer. . . ."

Trace amounts of it are found throughout the world in the environment, but dioxin can rapidly accumulate in the food chain, and large volumes of artificially introduced sources of dioxin, such as an emergency drop of a planeload of Agent Orange in an airfield, have the potential to cause higher mammals to be inflicted with doses that are dozens of times higher than the background exposure at that geographic location. Under those circumstances, this class of compounds can be highly toxic.

But what about now, countrywide, nearly forty years after the last spraying was done?

In some ways, the long-term issue of dioxin in the spraying of Agent Orange can be compared to that of low levels of background radiation still present in Hiroshima and Nagasaki decades after the bombing of those cities at the end of World War II. Both actions occurred during wartime, before anyone knew just how dangerous these unintended components of the bombing were to humans, and it was only after years of work that scientists had some idea of their long-term impact on the environment, plants, animals, and humans.

What we do know about the long-term effects of lower levels of radiation has come about through studies aimed, logically enough, on the impact on humans, in which animals with similar systems such as lab mice were exposed to very large volumes of radiation. Consequently, we have a pretty clear idea of just how bad things are—increased levels of birth defects, mutations, cancers—when any creature is exposed to large amounts of radiation in a short period of time. Things get more confused when dealing with prolonged exposure to low levels of radiation over a long period of time. And to complicate matters, there is a certain amount of background exposure that happens naturally from small, everyday sources—cosmic rays from space, television sets, and even flying in a plane at high altitude can give one a much higher dose of radiation than is normally found in the background.

The same is true of dioxin, in the sense that a small amount is always present in the environment. In fact, you could replace the word "radiation" with "dioxin" in the above paragraph, and that would sum up the situation with prolonged exposure to dioxin in the environment over time, be it to fish, bird, animal, or human.

To find out what the effect of dioxin is upon wildlife, we have to take all these matters into account and extrapolate from what studies have been done upon humans. Other than studies such as Martha Hurley's on the persistence of dioxin in the overall environment of Vietnam, there have been few or no studies on the impact of Agent Orange and its key contaminant, dioxin, on specific wildlife. (That being said, it can be hard to prove a negative and say that absolutely no studies have been done anywhere on, say, the effects of dioxin on the saola today.) As of this writing, I could find no systematic, long-term data on the effects of dioxin on Vietnamese wild fauna.

For now, the best science can do is say that some of the dioxin is still there in the natural environment, still bioaccumulating in the fatty tissues of wildlife as it works its way up the food chain. We know that at high levels, dioxin contributes to birth defects, cancers, skin diseases, and other conditions, but we are still working out its effects at low levels over longer time periods. Dioxin is still being found in Vietnam's freshwater creatures, decades after the war ended.

In addition to the effects of the dioxin it contained, Agent Orange certainly was an aggressive defoliant, reducing the amount of natural habitat available to langurs, gibbons, civets, deer, bears, gaur, saolas, koupreys, leopards, cranes, and many other species of animals, stripping the forest of its native plant species, increasing erosion, and allowing foreign exotics to gain a foothold. Repeated applications of Agent Orange—and the rainbow of other defoliants—had sometimes eradicated all native vegetation, and the natural environment of Vietnam has still not recovered to its previous state in many places.

Beyond that, only time and further study will tell.

"The confusion of the war lives on," said Stellman, the Agent Orange expert.

12

DMZ: The Thin Green Line

Animals don't need visas.

—Vo Quy, dean of biology, Hanoi University

SO HOW DID Vietnam's wildlife manage to not only survive the Vietnam/American War but also to remain unseen while doing it? Besides Agent Orange, how did they elude B-52 bombing raids, fire-fights, and minefields? Especially at a time when large creatures such as Asian elephants (*Elephas maximus*) were deliberately napalmed by Air Force pilots in order to deprive the Viet Cong—the Communist North's irregular forces fighting against the American-supported South—of a time-honored way of transporting heavy war matériel?

Understanding how the wildlife beat the odds is important, because the next "biological gold rush" will probably be in similarly closed-off, tropical locales such as Cuba, the "hermit kingdom" of North Korea, or Myanmar (Burma)—anyplace currently off-limits.

Some answers come from a man who served on the opposing side during the war: Professor Vo Quy is the retired dean of biology at Hanoi University, and former chairman of a state commission charged with protecting natural resources. He wrote the first Vietnamese-language bird guide, the two-volume *Birds of Vietnam;* founded the country's first environmental research institute in 1954; and had a hand in establishing its nature reserves. He was the architect behind the "National Conservation Strategy," which mobilized the Vietnamese people to plant nearly 400,000 acres of trees per year to make up for the loss of millions of acres of forest and farmland damaged during the war, and was later awarded the World Wildlife Fund's Gold Medal for his work. The grandfatherly scientist was *Time* magazine's "Hero of the Environment" in 2008, and they wrote of him: "He protected Vietnam's flora and fauna from the ravages of war. Now he fears the toll of torrid economic growth."

Fluent in five languages, the white-haired, bright-eyed, diminutive Vo Quy was at the center of a number of legends, some of which were true: Vo Quy personally lobbied Ho Chi Minh to establish Cuc Phuong in 1962 as Vietnam's first national park; in the early 1970s, he led an expedition into a war zone to do the mammal equivalent of a bird-watching trip. Later, he studied the toxic legacy of Agent Orange. But beneath his gentle, bemused exterior, Vo Quy gives the impression of a very determined and persuasive individual. As someone who has survived in the Communist government hierarchy for decades, he must have been a skilled political infighter as well.

Our interview took place at his Hanoi office. Antlers, skins, and animal posters covered the walls; monographs were stacked in piles on every square millimeter of flat surface; and a complete adult elephant skeleton stood in the high-ceilinged hall outside his door. An old mahogany desk and a slowly rotating ceiling fan completed the scene.

Even though he was an articulate English-speaker, we were still required to have an officially approved, Party-mandated "interpreter" at all times, as per the agreement with the people who had arranged my visa during this visit. Despite the interpreter's presence, and the knowledge that whatever words Vo Quy uttered would be reported back to someone, Vo Quy did not appear to hold anything back and at times was quite critical of the government for being slow in putting teeth into protecting the parks and preserves he had so laboriously assembled. It may be that as the grand old man of Vietnamese biology, well known in the West, with ties back to the earliest days of the Party, Vo Quy was beyond reproach. And surprisingly, our government minder turned out to be a low-key, pleasant, and capable organizer—actually less of a problem than some corporate public-relations flacks—with the good sense to stay in the background as much as possible.

Vo Quy noted that Vietnam's newly discovered species survived the war partly because of the country's sheer size: At 127,240 square miles, Vietnam has a landmass about three-quarters the size of California, or a little smaller than Germany. It also has about two thousand miles of coastline, and about three thousand islands—where many species have found a refuge, far away from whatever happened on the mainland. Indeed, one island, known as "Con Dao" or "Poulo Condore" and formerly the Devil's Island of Indochina, is now a sea turtle sanctuary; nearly 80 percent of the island's original vegetation is still intact.

Given the country's immensity, for logistical reasons alone, only "high-value targets" were hit by the US military, with much of the bombing confined to places such as the Ho Chi Minh Trail, the border between North and South Vietnam, or known Viet Cong bases. Consequently, large amounts of the countryside escaped direct hits. (This is not to say that there were not humanitarian concerns as well. For example, the reservoirs serving Hanoi were never attacked, for fear of drowning large numbers of innocent civilians.)

This meant that although more than 850 square miles of forest were destroyed during the Second Indochina War, the vast majority was spared. Most animal habitat was far from combat. And the many fissures and folds in the steep terrain of wet, old-growth forest—the preferred environment of newly found rare mammals such as the saola—proved difficult to bomb effectively. Vo Quy was sure that many other creatures similarly managed to escape, and he was convinced they are still out there. "The local people are always finding things that we scientists do not know about," he said—a statement that was supported later by outsiders such as biologist George Schaller. ("In Vietnam, I once ate a pig for dinner which had been thought extinct since 1892," said Schaller. Some measurements by taxonomist Colin Groves later confirmed that it was indeed the "missing" animal.)

Vo Quy cited a wartime experience of his own to explain how wildlife eluded destruction. A man of the Northern, Communist side, Vo Quy was leading a group of nine people in a wildlife survey in the early 1970s along the Ho Chi Minh Trail—the supply line between the rear base in Hanoi and the front-line troops in the south—when it became too dark and rainy to continue. Members of the group set up their tarps and hammocks in total darkness. When they awoke the next morning, they found that the earth all around them was covered with "butterfly" mines: small, winged antipersonnel mines that had been dropped from the air some time earlier. Fortunately, the monsoon rains had caused the mines to rust, while softening the ground so much that these pressure-sensitive mines failed to go off. He smiled at the thought: none of them had wanted to get out of their hammocks that morning.

Vietnam's myriad caves also helped preserve wildlife. Many animals had access to such caves, usually located in narrow, deep, forgotten, jungle-covered valleys that were essentially out of the way of anything but a direct hit. With some caves big enough to swallow entire city blocks, the caves provided ample space in which to hide.

Going underground is a well-known strategy   and why so many elaborate underground bunker complexes were dug in Vietnam. A person gets a real appreciation for these man-made caves when visiting places like the tunnels of Vinh Moc in the old fought-over zone on the coast near Quang Tri, where the former North and South Vietnam met.

The complex had thousands of yards of tunnels, divided into three levels, with thirteen separate entrances, along with a hidden exit to the beach. Dug by hand into the red clay, the bunker complex allowed three hundred people to live there for weeks at a time, hiding by day and coming out at night to till the fields.

One can only imagine what it would have been like to live for weeks at a time in this narrow, confined space, about a hundred feet belowground. But visiting the Vinh Moc tunnels showed how effective a haven an underground space can be for people during even the heaviest B-52 bombing campaign. And how wildlife could similarly shelter belowground.

What's more, in a bizarre twist, the peculiarities of warfare may actually have helped to protect the wildlife. Combat so preoccupied both sides that all of their resources and effort went into fighting, leaving little money for road building or development—activities that have long-term effects on habitat. Likewise, few people wanted to venture anywhere near the site of a battlefield for long afterward. It was just too risky for hunting or setting snare lines—the latter a destructive practice in which fences of brush with snares at intervals extend for hundreds of yards over slopes and ridges, and indiscriminately catch small- to medium-sized animals, and which some have referred to as the "drift nets of the land." "Many times, hunters don't even check the snares regularly to see what they've caught; they might only check once every three weeks," said Nicholas Wilkinson of Cambridge University's Darwin Initiative Project.

Many villagers are still afraid to go into the depths of bomb-strewn old battle zones, and the trees are so full of shrapnel that they are worthless to lumber mills. The observation that low-key, steady warfare can actually be a boon to wildlife was well known to wildlife researchers—although it has to be the right kind of hostilities, where the situation is stable, such as demilitarized zones. "If you go to Afghanistan, where everyone carries an automatic rifle, that's not very healthy for wildlife," commented George Schaller.

Another reason war may have inadvertently protected wildlife may be what researchers such as the ANU's Colin Groves call the "border effect." Simply put, boundary areas between nations make for effective wildlife sanctuaries, as long as there are no active battles going on. Few people want to provoke trouble by going into no-man's-land. Consequences can be dire: in the area where Laos and Vietnam converge—called the Parrot's Beak for its shape on the map—a Vietnamese wildlife researcher was shot by Cambodian Khmer Rouge guerrillas while trying to find new species.

Meanwhile, animal life pays no attention to human-drawn political boundaries. "Vietnam shares a lot of species and a lot of habitat with Laos and with Cambodia, so a lot of these endemic animals live there, especially along the Annamese Cordillera," says Martha Hurley of the American Museum of Natural History in New York. "They don't care about the border."

Or as Vo Quy put it: "Animals don't need visas."

The border effect may have been around for a long time, in many places, right under our noses. When Lewis and Clark did their great exploration of the North American continent, they noted much more wild game in the buffers between Indian nations. Analyzing the explorers' old journals today, researchers Paul Martin and Christine Szuter found that the border areas were avoided by the native tribes of the Lewis and Clark era. Unless they were actively going to war, these tribes deliberately avoided the boundary lands along the upper Missouri and Yellowstone Rivers that marked the edges of their respective turfs; the explorers' notes show much larger numbers of bison, elk, deer, pronghorn antelope, and wolves in these areas.

Similarly, Joseph Dudley and his colleagues wrote in a 2002 study in *Conservation Biology*, "Effects of War and Civil Strife on Wildlife and Wildlife Habitats," that no-man's-lands have long served as de facto nature reserves. One of the study's coauthors, Andrew Plumptre, told a science magazine that the bottom line seems to be that species bounce back when humans leave them and their habitat alone for prolonged periods. "War can be good in that it keeps people from moving into an area and settling there."

Or, as Vo Quy said, "In some ways, the peace is more dangerous than war for endangered species." The border effect shows up worldwide, in unexpected places, both large and small. England, for example, is renowned for the many hedgerows that mark the boundaries of its farmers' fields; in the mid–twentieth century there were an estimated 500,000 miles of hedgerow. Up to ten feet wide, these strips of land between fields and pastures cumulatively make up a huge square footage of land area; the hedgerows form a wild nature reserve in the midst of an otherwise heavily domesticated island. Often overlooked, hedgerows are now seen as more than mere boundary markers but as an important source of biodiversity. For that reason, there are active campaigns, such as the Hedgerow Trust, to preserve them today.

If there is that much biodiversity going on in these ten-foot-wide dividers between farmers' fields, imagine what it is like in the 2.5-mile-wide, 155-mile-long demilitarized zone (DMZ) that divides the two Koreas. The zone was established at the end of the three-year-long Korean War in 1953, and it has lain largely undisturbed ever since. So, while intensive industrialization and commercial agriculture have taken place elsewhere on the Korean Peninsula for more than half a century, this ribbon of land—which some have dubbed "the thin green line"—has remained untouched.

Left to its own devices, nature has made a comeback along the Thirty-eighth Parallel: about twenty thousand migratory birds come to use this border area each year, including the world's largest population of endangered red-crowned cranes (*Grus japonensis*), which nest long-term in this no-man's-land between two million soldiers from opposing armies. Cranes in general are spectacular species, as much as five feet high, with elaborate courtship dances and memorable calls. In describing the whooping crane (*Grus americana*), pioneering ecologist Aldo Leopold wrote, "When we hear his call we hear no mere bird. We hear the trumpet in the orchestra of evolution."

As a result of this inadvertent wildlife sanctuary, their population jumped from three hundred to eight hundred. There are also indications that the Asian black bear (*Ursus thibetanus*) and the Siberian musk deer (*Moschus moschiferus*) survive amid the tank traps, land mines, and tunnels of Korea's demilitarized zone.

The DMZ had effectively become an unintended nature reserve, where wild species could thrive—especially cranes, a cultural treasure on the Korean peninsula and a symbol of long life and good luck, said George Archibald, cofounder of the International Crane Foundation (ICF). The ICF was started in 1973 in Baraboo, Wisconsin, by Archibald and a biologist friend, Ron Sauey. Self-described as "two college kids on a farm who wanted to save a nearly extinct bird no one knew much about," the organization now has forty people on staff and focuses on fifteen different crane species, of which eleven species are classified as threatened. In many ways, Archibald's work at the ICF with cranes reminds me of Nadler's work at the EPRC with langurs.

Archibald described his work in the Korean DMZ a few days after returning from his latest conservation effort there; he had left North Korea less than thirty minutes before officials announced that Supreme Leader Kim Jong-il had died. Archibald had gone there to work with biologists at the Academy of Science in Pyongyang and had come up with ideas for ensuring the long-term success of all the different species in the DMZ. In general, any flat, arable land is very scarce in the North, leaving only about 5 percent of the country suitable for farming—and both cranes and people desire the same bit of land. With any kind of space being scarce in the Koreas, he expects that if the DMZ ever comes down, "It will all become a city overnight."

So Archibald is trying to find ways to prepare in advance to protect the land on which the DMZ sits—the same kind of advance planning that was done before the former East Germany united with West Germany, and why wildlife conservation was so successful there. He hopes to repeat the experience in Asia: "What will happen when and if the two Koreas unite is problematic at the moment," Archibald said, adding that working in geopolitical hot spots and dealing with potentially hostile countries was all part and parcel of the job.

His attitude is echoed by Alan Rabinowitz, a zoologist with the Bronx Zoo who is now the CEO of Panthera, an organization devoted to saving big cats such as jaguars, clouded leopards, Asiatic leopards, tigers, and other species. Rabinowitz says that he's found that sometimes it's actually easier to deal with dictatorships: "This is not a political statement, but the fact is that I and others in my field find that sometimes conservation is easier to do in communist countries and in dictatorships than in democracies." He says that with these top-down structures, there's only one person you have to sway, and that one person can order a change and it will be done. "They're not nice people, and I'm not an apologist for the ones I work with, but I will do anything I can to save the animals," Rabinowitz explained. "If I get to the right dictator and I can convince him that he should do something—and he can guarantee that he'll save tigers—then I'll do it."

In Archibald's case, dealing with a potentially unstable, nuclear-armed dictatorship sometimes got hairy. He quickly learned to be extremely diplomatic at all times: Archibald does not take pictures of anything that is off-limits, for example, and he tries to be sensitive to the political situation, doing benign projects such as showing farmers new agricultural techniques that can boost their yields, thus keeping starving people from having to resort to killing the birds. "My North Korean sources claim that no one has ever shot and eaten a crane itself . . . but I wonder," he said. "If someone is starving, I could understand how it could happen."

What's more, with the severe food shortages that the country has been having lately, he observed that many North Koreans have become desperate, digging up roots, shoots, weeds, aquatic plants, and any stray kernel that they can find in a field—so there was nothing left for cranes to feed on. But with better yields, he's seen the number of cranes go up by a factor of six in some areas, such as on the Anbyon Plain, a delta located away from the border.

To cover his bets, Archibald hopes to create sanctuaries in Korea outside of the DMZ. And the ICF has been working to bring back cranes in places such as Vietnam's Plain of Reeds, sending some of the first American biologists to Vietnam since war's end.

Asked why he does it, Archibald replies, "Actually, journalists are the only ones who ever ask me that. People realize that endangered species and endangered places have special value. You have museums to protect the works of man—which were created over a short span of time, in the big picture. These species were created over millions of years by nature; they're something rare and special to us. So, of course we should protect them."

He went on: "But I've never had to spell out the case like that to ordinary people. It's always journalists who ask me why we should save cranes."

Meanwhile, he is thankful for the presence of the demilitarized zone, which gives him some breathing space in which to work, and some possible land for the long term.

Said Archibald, "If it wasn't for the DMZ, there would be no cranes in Korea."

13

Dragons Flying in Clouds

The current destruction of our forests will lead to serious effects on climate, productivity, and life. The forest is gold. If we know how to conserve and manage it well, it will be very valuable.

—Ho Chi Minh, 1963

WE ARE ABOARD a pair of small square-bowed, flat-bottomed boats made out of bamboo and reed, being poled along the clear, clean, shallow waters of a lake about two hours from Hanoi. Somewhere above our heads, the Scratching Cat and the Lady Fairy go by.

The Cat and the Fairy are the names of just a few of the many skyscraper-like limestone karst blocks that jut up from the middle of the waters of Van Long Nature Reserve to form steep, near-vertical islands.

The tabletop-flat plain on which the lake sits is generously sprinkled with such mountainous islands, which make up in raw vertical height what they lack in width. Some karsts appear to be only about as wide as a couple of suburban ranch houses but tower nearly six hundred feet above the waterline. Some have truly odd shapes, looking like weirdly sculpted, enormous works of modern art that would not be out of place in a Salvador Dalí museum. Many seem to have razor-sharp tops, and slopes of sixty to eighty degrees are not uncommon, with faces composed of a series of stepped-back cliff tops and terraces that form an almost continuous network of narrow ledges overgrown with bushes, grasses, and the odd tree.

Each island has an intriguing name, including Orphan and Book. The names are rooted in old Vietnamese folktales from the region. Even "Van Long" is colorful: one source says it means "dragons flying in clouds."

The reserve is also home to three dozen caves, with similarly colorful monikers. If I had to vote for my favorite cave name, it would be Creep, so-called because of what people must do to avoid the stalactites dangling from the ceiling.

I had long been looking forward to visiting this spot with Nadler, which he had told me about in his office and over dinner at his home at the EPRC. It was a chance to see the place where their latest, most ambitious initiative was taking place: actually releasing into the wild some of the rarest and most endangered animals in the world. This same locale was the one that Hien thought so highly of; surprisingly, it did not appear on any online sources.

Located about halfway between the Endangered Primate Rescue Center (EPRC) and the provincial capital at Ninh Binh, Van Long's limestone islands are full of nooks, crannies, and hiding places that make for a wonderful habitat—the food, shelter, and other environmental factors that an organism needs to survive. This in turn makes Van Long a prime place for visitors to see all kinds of rare wildlife in its natural environment.

As an added bonus, any animal here is much easier to spot on these relatively bare cliff sides than in the deep, quintuple-canopy jungle outside the Van Long Reserve; at least one zoological report says tersely that the limestone forest is "a most strenuous habitat to walk," which is a bit of an understatement.

And which is why I am here on a boat, and why Tilo Nadler and one of his colleagues are in another one about forty yards away. We're here to look for what may be the only viable population of wild Delacour's langurs left in the world.

The International Union for the Conservation of Nature puts the Delacour's langur, or *Trachypithecus delacouri,* on its list of "The World's 25 Most Endangered Primates" closest to extinction. (Primates are mammals that have front-facing eyes, nails instead of claws, and hands and feet with the ability to grasp objects.) Depending on whom you talk to, there are about two hundred Delacour's langurs left on the entire planet, and Nadler's best estimate is that about a hundred of these are here in Van Long. In biological terms, these tiny numbers mean that they are on the very edge of extinction—a knife edge as sharp as the top of one of the island peaks they inhabit.

The animal, named after one Jean Delacour, a prominent ornithologist who was one of the few explorers to collect mammal specimens in what was then the "Union of French Indochina," was first written up in the scientific literature in the early 1930s. In the time since then, "there was only scanty information on its existence and distribution," says an IUCN report, which adds, "The most important and for some subpopulations the only factor for the decline in numbers is poaching, which is not primarily for meat, but for bones, organs and tissues that are used in the preparation of traditional medicines."

That particular IUCN entry, it turns out, was written by Tilo Nadler. And it sums up the ten years of surveys that he has done for the Frankfurt Zoological Society very starkly: only about 280 animals living in the year 1999, with another 320 killed in the ensuing decade, for an annual loss of more than thirty individuals. In other words, roughly as fast as they bear young, they are killed off.

Most of those that survive live in small, isolated populations of fewer than twenty animals each—far too few in each little cluster for a viable breeding population. Most likely, each of these little bunches of animals does not have enough genetic variety by itself to survive in isolation over the long haul.

But here in Van Long, the picture is much brighter. The population of Delacour's langurs at this site has actually increased, more than doubling in the past few years. When I related this information to other zoologists, they were pleased by the unexpected good news, and remarked that despite Vietnam's reputation as a place where conservation laws only existed on paper, they could see progress. George Schaller commented, "It shows that if you focus on specific species in specific areas in specific countries, you can make a definite improvement. The Chinese are doing better on protecting Tibetan antelope, Brazil is improving in its protection of jaguars, Rwanda is working very hard to protect its mountain gorillas—and their numbers are increasing. . . . So you can point to some progress, which is important."

Alan Rabinowitz agreed, saying he saw hope for specific animals in specific places with targeted programs, especially if they enlisted the local community. Rabinowitz added, "I think the key to a good protection program is for someone to be passionate. The money doesn't need to go to incredibly expensive structures; you can start with a great model and scale things up. But none of that happens without passion . . . and that passion can inspire others."

In line with this idea, the landmass of the Van Long reserve has increased in size. Soon, it may be a formally recognized national park in itself, giving hope for the long-term future of these animals and their endangered kin.

Meanwhile, this nature preserve has strong wildlife protections; some of them come from the top down, but most are from bottom-up initiatives. The leaders of the local commune were the ones who initiated a ban on hunting in the reserve, and they also provided the impetus to eradicate domestic goats and other animals that were killing all the wild vegetation. These regulations are enforced by a combination of the efforts of the local provincial authorities and a special ranger patrol paid for by the FZS. Saving this species has been a primary focus of the FZS and of Nadler's Endangered Primate Rescue Center, and although Nadler himself may be too self-effacing to say it, I got the impression from other researchers that the patrols would not exist at all if not for his efforts. The FZS and the EPRC work closely with the management board of Van Long Nature Reserve, helping to survey and mark its borders, construct five ranger stations, pay wages to its staff, and arrange for uniforms and equipment for twenty guards. (The EPRC has also helped to establish a similar, though much smaller, "release site" for Ha Tinh langurs in a portion of the 330-square-mile Phong Nha-Ke Bang National Park farther south, in Central Vietnam next to the Laos border. It boasts a limestone-based environment roughly analogous to that of Van Long, minus the lake.)

The EPRC began its captive breeding program with five Delacour's langurs rescued from poachers, and fifteen individuals of just this species alone have been born since the EPRC was established in 1993—in addition to all the other endangered animals the organization has rescued from the very brink.

But the EPRC itself was really just a temporary holding pen, or "species reserve" for animals prior to being returned to the wild, which is where a much larger space like Van Long comes into play. The langurs cannot stay in captivity forever, as they would lose their vital survival skills and become dependent on daily handouts. Nor could they just be dumped at random in the woods—"There's no point in keeping them alive at the EPRC unless you can return them to a protected area in the wild. Otherwise, they'd just be caught again or killed," explained Nadler.

The influence of the Van Long Reserve reflects the changing aims of the EPRC over the years. At first, the EPRC's goal was only to rescue single animals of the very rarest and most endangered species from immediate danger. It was really sort of a temporary holding pen.

But over time, they realized that these animals didn't have a place to go to after rehabilitation; the creatures had no chance of surviving in the wild because of the extremely high hunting pressure. "There was no place to release them to," said Nadler simply. "At the same time, we began getting more and more animals confiscated from the illegal trade. Nobody had any expectation how many animals we would get. And in doing surveys of the area, I found out that the wild populations were decreasing very fast. That was the background for trying to establish small but viable populations in captivity as a reservoir to strengthen or reestablish wild populations in their original habitats."

The idea behind this strategy was that this effort would occur in parallel with other projects—also in cooperation with Vietnamese institutions and foreign organizations—that sought to preserve the natural environment and provide sustainable ways for local people to earn a living. The hope was that by all these things working in tandem, they could together reduce hunting pressure.

Nadler elaborated: "Hopefully, we can rescue both the species and their natural habitats. For both steps—reducing hunting pressure and establishing small, stable populations in captivity—time is necessary. And only if we have both does it make sense to release animals into the wild." The semiwild areas at the EPRC were the first step, allowing them to learn what these previously unknown animals needed to survive, in a somewhat controlled and safe environment, at an acceptable level of risk.

The next step was releasing them into protected large-scale, open-air wildlife preserves such as Van Long, which were free of poachers.

Those first few days afterward are always nerve-wracking, in which a lot of fingers are crossed as to how the animals are coping with the situation in the wild after being kept in captivity or born in a cage. To keep tabs on the animals after release, some individuals are released with a radio collar, and then the researchers hold their collective breath. If the collars tell them that there is movement, that's a good sign. There's nothing worse than when a radio collar never moves again.

To me, it seemed that the role of the EPRC could be thought of as something like the intensive-care unit of a hospital, in which a small population of the most critically endangered animals is placed under close, round-the-clock monitoring and supervision while their health improves. In the EPRC's case, the staff also cares for the young offspring, until the species' numbers are brought up to something resembling stability.

For their part, the parks, reserves, and other sanctuaries act more as the general ward for the population once they're out of intensive care, where the animals can start to fend for themselves and begin to lead more fully independent lives while still free from the threats posed by hunters. Hopefully, at some point the animals' population thrives enough so that they can spread out from there into the countryside at large, reclaiming their former range.

And so Van Long came to pass, after much work with the Vietnamese government and coordination with outside organizations such as a Dutch non-profit. By protecting this patch of prime habitat and reintroducing the captive-bred population into an existing, healthy, wild-ranging native population, the finicky leaf-eating langurs could have a chance at staging a comeback.

Early signs, admittedly tentative, are promising. In an interview at the EPRC headquarters the day before I visited Van Long, Nadler told me, "It used to be that you would see primates at Van Long once out of every twenty field visits. Now things are reversed, and you are just about guaranteed to see several of them on each trip. . . . It shows what you can do if you focus on one area."

What's more, the langurs you do see in the jungle are behaving more like wild langurs are supposed to. There are no artificial behaviors, nor the eating of the langur equivalent of junk food. It's similar to what happened when Yellowstone National Park closed its trash dumps years ago, forcing large communal groups of "garbage bears" to stop rummaging in the waste for edibles. When I worked there as a volunteer during the summer months in college, the rangers told us that with the dumps closed, the bears were no longer able to get "food" by licking the insides of jelly jars or swallowing Twinkies wrappers, and the animals began going back to eating more natural materials such as berries and wild game, which was better for their overall health and more in line with the requirements of their dietary system. And once the dumps were closed, the bears started behaving in ways that were more truly "bearlike"—living more independent and solitary lives in the wilderness, no longer scouring garbage dumps or begging by the roadside for handouts in unnaturally large groups, and consequently having fewer bad close encounters with human tourists.

These human-bear interactions could be ugly, and sometimes had the potential for truly awful results. The old-timers among the rangers often cited the time that one visitor, observing the bears' fascination with jam, smeared the face of his five-year-old son with Smucker's grape jelly and stuck the kid out of the car window in order to get a photo of a grizzly bear, *Ursus arctos horribilis* (the scientific name literally means "horrible bear") licking his child's face. According to the lore, he was stopped in the nick of time, but could not understand why the rangers were so upset.

I had always dismissed this story as legend. But after reading Lee Whittlesey's *Death in Yellowstone: Accidents and Foolhardiness in the First National Park,* I began to give the jam story more credence. All of which convinced me that the wilderness and its creatures should be appreciated for what they are: wild and unpredictable, to be appreciated on their turf on their own terms.

So we hopped into an EPRC van and drove to Van Long Reserve, hoping to see wild Delacour's langurs—a leaf-eating creature that is in many ways a symbol of the EPRC organization—in their natural environment.

Our boatmen—actually older women in conical hats and the traditional loose blouses and trousers of country Vietnam—get into the spirit of the quest, and compete with one another to see who can be the first to spot langurs on the mountainsides. From what I can gather, the reserve has strong support from the local community: the steady, if not overwhelmingly large, flow of ecotourists brings in a decent supply of hard currency to supplement the locals' income. Villagers make money off the fees for paddling people around along with the occasional meal or snack, plus charging for whatever accommodations the tourists want.

There are non-monetary rewards as well: There is a tangible sense that the villagers seem proud of their nature preserve, from the impression I was able to get from my decidedly nonscientific encounters. Such positive local attitudes contrasted vividly to what happened just two days earlier in another part of Vietnam, when a mob of local people attacked a ranger station after the rangers had arrested some illegal timber cutters. Winning over "hearts and minds," to coin a phrase, still appears to matter in Vietnam. This impression was supported later by the comments of a wildlife researcher at the Center for Natural Resources and Environmental Studies in Hanoi, Nguyen Manh Ha. Nguyen was born and raised just outside Cuc Phuong, where his father had worked as a ranger for twenty-six years, and he considered Van Long to be a real success story when it came to local involvement. It was still an uphill battle, however, he said: "This is a poor country, so it can be hard to justify spending a lot of money on wildlife conservation to some of the local people."

The depth of the water in the reserve varies from chest-deep to barely over a foot, so the boatwomen constantly shift from paddling to poling and back. The only sound is the splash of wood on water or the shuffle of bare feet on bamboo. When the boats drift together

closely enough for conversation, Nadler joins in the women's ban-
ter, and I have the feeling that his Vietnamese-language skills must
be much better than he lets on: when we were negotiating the fee
for our boat rides on land earlier, he greeted many of the people
on the wharf by name. At least three people came up to chat, amid
much handshaking and laughing, and one slapped him convivially
on the back.

Then we spot the first Delacour's langur, halfway up the cliff
side. After a lot of pointing and gesticulating in a mixture of Vietnamese, German, and English, I can see the shy and cautious animal as well, by looking through some binoculars loaned to me. We
all observe the creature while it scoots along the gray ledges and
feeds upon the leaves of the bushes growing there. Later we spot
two more.

Their untamed reality makes you draw your breath, conjuring
up the poet Ted Hughes's phrase "wildernesses of freedom." It's
incredible to realize that there are very few of these animals in the
entire world, and you've just seen three of them in the wild, while
next to the person responsible for them being alive and well. Such
opportunities are one of the reasons why people become journalists.
"The freelance life can be hard, but it has its rewards," as a friend
used to say.

Peering intently, I can make out the glossy black of the legs and
the similarly black upper half of the body of the Delacour's langur,
along with its telltale pure-white waist and rump. The boundary
where white fur meets black is clearly defined, to form a straight
line across the animal's lower midsection. It is easy to understand
why local Vietnamese call the creature *vooc mong lang,* or "leaf monkey with white trousers."

From this distance, I cannot make out the animal's luxurious
long tail amid the cliff-side bushes, which is surprising given that
the ones I've seen in captivity seem to be almost all tail. In fact, I
overheard someone at the EPRC joke that these critters look more

like a big tail with a little langur attached. Curiously, these appendages cannot grasp anything—a langur's tail is not "prehensile" like
a spider monkey's is. Instead, langurs' tails are used only for balance.

It is amazing to think that I have seen this animal in its natural home, and I can only hope that I am not one of the last to
do so. One early explorer called the Delacour's langur a "hand-
some monkey" and described seeing one as a "conspicuous result"
of his expedition. Even from a distance, it is easy to see why the
animal got such praise. (Like many casual observers, this explorer
had confused leaf-eating langurs with their fruit-eating monkey
cousins—something that happens to this day and is responsible for
much confusion among those trying to feed and care for the creatures, a situation made all the more serious when the animals die
from being fed the wrong foods.)

Once you know what you are looking for, the creatures are fairly
easy to spot, and it becomes clear that our coracle-like boats are the
best way to overcome the defenses of the animals' natural moat, allowing us to approach closely enough to take pictures. From their
movements, I can see that these creatures are fussy about what they
eat, rejecting the leaves on one bush for a seemingly identical set of
leaves on another. It's a behavior familiar to anyone who has tended
domestic farm animals. The ability to make such discriminations
may allow grazing animals such as sheep or goats to avoid poisonous
foliage while seeking out the most nutritious, best-tasting greens.

As our vessels thread their way among the waterways, chan-
nels, sandbanks, and occasional banks of reeds, I notice that there
is hardly anyone else around. It may be that this solitude—rare in
Vietnam—is due to the light mist overhead, turning this Sunday
afternoon into an exceptionally slow day.

Or perhaps the uncrowded feeling is because at 3,000 hectares
(approximately 7,400 acres, or 12 square miles), this freshwater wetland preserve is large enough to have plenty of room while still being
small enough to remain undiscovered by commercial mass tourism.

With so few visitors and so much land, there's not much chance of running into others, and the place is still largely unknown. Van Long Nature Reserve doesn't appear on Google Maps and there is scant information about it in any guidebook or on Wikipedia. But considering that the concrete foundations of a "5-Star Van Long Eco Tourism Resort Complex" are being poured close to the village wharf, all that may change soon. At times, I worry whether I am helping to bring on the place's downfall by the very act of writing about it.

For the moment, at least, floating in Van Long on a cool, lightly overcast day in the spring has to be one of the most relaxing ways to look for Asian wildlife in the bush, especially after dealing with the frenzy of a city in a developing country. As an added benefit, due to this wildlife reserve being so far off the beaten path, it was never a military target, so there is no need to be concerned about unexploded ordnance. And because I am safely inside a boat, no worries about leeches. In addition, because the reserve is composed of moving water instead of stagnant pools, there are not even any mosquitoes, and the open air feels refreshing after the closeness of the jungle of Cuc Phuong.

A flock of white storks goes by, along with several kinds of egrets, which number about two thousand at the park and comprise at least seven different species. This preserve is also well known as a haven for migrating raptors, and a biology reference guide formally known as the *Sourcebook of Existing and Proposed Protected Areas in Vietnam* (or more popularly as the "Vietnam Source Book") says that Van Long is the only place in the country to see the rare Bonelli's eagle, *Hieraaetus fasciatus*.

This sensation of being in another world is increased when a small shrine appears, tucked in among the water-level caves and fissures of the shoreline. Shrines and pagodas are sprinkled abundantly around the reserve, some dating back to the tenth century, when this area was home to the rulers of the Le dynasty.

Meanwhile, Tilo Nadler is busy on the other boat taking photos with a three-foot-long telephoto lens. The camera equipment looks as long as a baseball bat, but with its tremendous magnification, this heavy piece of optical gear brings the animals in much closer. From this distance, with his green-gray uniform and EPRC shoulder patch, Nadler looks like an overwrought statue of "Ranger on Patrol," even if that may be more in title than in fact, as he has no formal authority here.

Just at that moment, there are excited shouts from the women poling our boats. They have spotted a poacher doing "electric fishing." With the use of a car battery on his back and a set of copper wires, the poacher is wading in the water, electrocuting anything he sees.

Though it started out as a more or less benign research tool, electric fishing can be an indiscriminate process in the wrong hands, killing all organisms within range, not just the ones the fisherman is looking for. To put it mildly, it is surprising to see this happening in broad daylight within a wildlife sanctuary. Such practices are strictly forbidden here, and carry a fine of five million Vietnamese dong, or about $240—a fortune in a place where most people get by on the equivalent of $1,100 per year.

Nadler's response is immediate.

He takes off his sandals, removes his pants, and wades in after the poacher. Clad only in his underwear and a shirt, and carrying the bulky camera with its telephoto lens, the sixty-eight-year-old Nadler chases after the poacher, who steadily edges away at high speed—or as fast as one can when submerged waist-deep in a lake, feet sinking into the mud at every step.

Though Nadler does not catch up with the poacher, he is still satisfied when he returns to his boat and gets dressed. "I got him on camera," Nadler shouts across the water, explaining that he will give the images to the authorities so that they will have evidence in hand for use in capturing the poacher and bringing him to trial.

Nadler takes a tough line on enforcing the law when it comes to environmental protection, and this attitude had much to do with the ranger training program. On the drive back from Van Long in an EPRC vehicle, he explains that when it comes to creatures on the brink of extinction, there is not much leeway. "In Vietnam, we have really good laws and regulations compared to other countries in the region. If we could just enforce them, we'd have a chance. But if they're not enforced, then what's the use? What's the point of making agreements about protected areas if no one obeys the rules?"

Because of Nadler's attitude about order, rules, and regulations, I detect that he is still a Northern European at heart, despite his two decades in Vietnam. For the last three years, as of the time of my latest visit, I had been working as a science writer in Switzerland, where in some cantons, policemen carried rulers to make sure that your car was parked exactly the proper distance from the curb—no more and no less. And if you lived in an apartment building, it was against the law to flush the toilet between the hours of ten p.m. and seven a.m. because the sound might disturb your neighbors. Similarly, each tenant had an assigned day and hour for doing laundry in the building's machines; if you're the newest arrival, you might be stuck with mid-morning on a weekday. The topper, however, was when a friend in Lausanne was stopped one day on her way to work because her boots "made too much noise on the sidewalk."

So I initially took Nadler's concern about rules with a grain of salt. But many zoologists subsequently said that Vietnam simply doesn't follow its own regulations, and instead acts as a major wild-life trading center. Some of the laws they do have are "outrageous," in which animals confiscated from the black market are often sold to the public—sometimes to the very same poachers hauled into court.

Nadler says that he feels that the average NGO is too timid when it comes to helping to enforce the wildlife-protection laws already on the books. To his mind, a stick is needed just as often as a carrot, especially when dealing with harmful activities happening within defined, clearly sign-posted and protected areas. In the bigger picture, there's always a battle over strategy. On the one hand, there are the social programs, sustainable development, and people-oriented projects that NGOs love, which have great fund-raising appeal. On the other hand, these programs drain time, money, and energy from the original focus: the unrelenting, systematic patrolling and protection of core habitat and its wild animals. "Too little is spent on the nuts and bolts of ranger work, such as jeeps, equipment, or guns. No NGO wants to buy pistols; they just want to buy posters," Nadler declared. It's a topic that we return to often as he points out how his rangers can be hamstrung by excessive, contradictory regulations and oversight.

It's a common concern among field biologists. In an ideal world, you could have it all: protection of core sanctuaries, healthy breeding populations in the wild, wildlife corridors so animals can move freely from one sanctuary to another, and small-scale, sustainable development of the surrounding human communities—but not too close to the core. In reality, there's always going to be more emphasis on one aspect or another, from different non-profits.

It was partly because of such problems that Nadler merely took photographs of the poacher. Rangers in Vietnam generally suffer from low status and low pay, below that of members of the Army or police. (For that matter, back home in the United States, rangers are often shockingly ill-paid as well. A ranger friend is hired every spring by the US National Park Service and handed a set of unemployment forms every autumn, with the same farewell each time: "See you next year." This has been going on for years. "Mostly, I get paid in sunsets," he said with a sigh.) A lot of even the most elemental necessities are in short supply. In addition to vehicles, Vietnam's rangers are often lacking things such as binoculars, GPS units, and communications equipment. What they do have is under tight constrictions. "They have a gun but are not allowed to shoot," said Nadler.

The rangers in Vietnam can temporarily hold a suspect but cannot make arrests, and they can only stop a vehicle if they know for a fact that there is illegal wildlife or smuggled timber aboard. Hence the benefit of photographic proof: the rangers at Van Long Reserve now have the materials they need to seize this particular poacher the next time they see him, and armed with this evidence, arrange for the police to come.

Similarly, rangers at Cuc Phuong National Park cannot simply take the car that Nadler bought for them and go. Instead, they must petition the park authorities for permission to use it, and fill out a form justifying the usage. They can no longer just patrol.

It's all very tedious and roundabout, partly due to bureaucracy and partly to battles for turf among the various authorities in the Vietnamese government. The most dramatic example of this occurred about five years ago, when bulldozers showed up at the northern section of Cuc Phuong National Park, ramming a portion of a new superhighway right through the top quarter of the park, effectively isolating it from the rest of Cuc Phuong. There was no prior notice from the Department of Science and Technology, which handles engineering and highways. They simply did not tell the Department of Agriculture and Rural Development, which runs the parks. The result was a standoff between construction crews and park rangers, which only ended when the rangers started showing up at the contested site with their rifles by their sides.

Construction was delayed.

During this time, the park's officials and about twenty NGOs signed a joint letter of protest and suggested an alternative, to no avail. After a year passed with no response from authorities, the bulldozers showed up again, and the park now remains split into two very unequal pieces.

"I could write a book about all the things that have been done wrong, or were poorly thought out or poorly implemented," declared Nadler, greatly animated. Some of the best, most ecologically aware Vietnamese administrators he has dealt with have been edged aside in favor of those seeking to promote headlong economic development regardless of environmental cost.

Nadler says that while he is wholeheartedly in favor of roads and economic development for local villagers, he feels that such projects should be carefully thought out so that they do not inadvertently encourage the destruction of the very thing they are meant to save. "Often, these projects are not sustainable at all, but just a masquerade," he said. "So-called win-win situations seem to always result in yet another power plant or road being built in fragile habitat that is home to endangered species."

Pointing out a former cement plant on the outskirts of the Van Long reserve, which used to demolish these limestone mountains to make the ingredients for concrete, Nadler says that there are small victories. Van Long is a good example of things working well, but "no one ever studies the success or failure of a given project, and why it occurred. They never want to learn from the past, but just want to start from scratch all the time, reinventing the wheel."

He must have been doing something right, because at least three other wildlife rescue projects had been spun off from the success of the Endangered Primate Rescue Center. And a good part of one rescue project was not even located in a jungle, but took place in one of the most populous parts of the entire country.

14

Of Turtles and
Second Chances

There's four left in the world and only one of them is female. You can't get more endangered than that without being hopeless.

—Paul Calle, Wildlife Conservation Society, to a reporter

DRAGONS SPIT FIRE at each other, boats race, a royal parade welcomes a scholar home to a village. A sea battle takes place, complete with ships that fire miniature cannons spewing puffs of smoke over a tiny sea. All the while, an orchestra of drums, Chinese guitars, bamboo flutes, and two-stringed violins plays in the background.

I am watching a performance of the Thang Long Water Puppet Theatre, located in a building in the center of downtown Hanoi, across the street from Hoan Kiem Lake. The theatre is home to the most prestigious of the country's fourteen best-known puppetry troupes; each night a show is put on that demonstrates the skills of the practitioners of this ancient craft.

A typical show consists of skits performed by as many as a dozen brightly painted, gilded, articulated wooden puppets, each about two feet or so high, floating on the surface of the dark waters of a shallow swimming pool. The puppets' legs and arms move, and they perform acrobatics in such a way that their movements seem very lifelike. All of their motions are controlled offstage by a team of up to eleven puppet masters as much as thirty feet away, hiding behind a screen and standing waist-deep in the inky water.

Beneath the surface of the water and hidden from view are long, thin strips of bamboo, along with pulleys and cables that allow the puppet masters to control everything their puppets do. (Despite their light weight, these bamboo poles can be extremely strong, as well as thin, flexible, and straight. "Tonkin cane bamboo," alias *Arundinaria amabilis* or *Pseudosasa amabilis,* is the stuff that premium, handmade fishing rods are made from; one company, Thomas and Thomas, charges as much as $4,600 for one of its bamboo rods, "built from the finest individually selected Tonkin cane." Hanoi sits on the Gulf of Tonkin.) Some puppets are so complicated that as many as three individuals are needed to operate just one of them.

Despite the fact that the whole performance is in Vietnamese, there are parts of the spectacle where a foreigner can tell exactly what is going on without needing any assistance, particularly during a skit depicting the courtship between two dragons. From the body language and the movement of the puppets, one can see that the girl dragon is flirting with the boy dragon, and that the boy dragon is becoming increasingly frustrated as he splashes around, chasing her.

Finally, after much back-and-forth, the two dragons kiss.

There is a pause and then a moment of complete silence, after which a baby dragon pops up from the water between them.

The skit ends with the male dragon proudly puffing away as the whole family parades around the pool before exiting. The whole performance may be a very stereotyped view of the traditional roles of the sexes, but perhaps because it is so stereotyped the skit reads instantly and overcomes any language barrier. The children in the audience scream with delight.

Found only in the northern half of Vietnam, water puppetry arose, some speculate, as a way for villagers to amuse themselves while using nothing more complex than colored bits of wood in murky farm ponds. The first written record of the practice dates back to the year 1121; over time, the performances grew more specialized and complex, with the knowledge passed down from father to son in highly regulated, very competitive guilds. Apprenticeships last a minimum of three years. There are rumors of at least one Westerner, an Englishman, who recently tried to apprentice himself and learn the craft.

Tradition holds that water puppetry was created by the gods, not the common people. As a result, each troupe has a small altar dedicated to the founders of the art form. Every year, troupe members organize a festival to honor the originators of their art; before each performance, they offer fruit, flowers, and incense to pray for a successful show.

In danger of fading out just a few decades ago, water puppetry has since made a roaring comeback and become a favorite of locals and tourists alike. The puppet shows are used to tell heroic myths, and one of the all-time favorites is skit number 12 on tonight's program. It takes place in the very lake whose shores lie just across the street from the front door of the theater building.

According to legend, when Vietnam was invaded by China's Ming dynasty early in the fifteenth century, a fisherman named Le Loi netted a magic sword in the lake's waters. Inscribed with the words "The Will of Heaven," the sword gave him the strength of many men.

With the sword's help, Le Loi became a general and waged a long war against the much larger Chinese army. By avoiding open battle, harassing and ambushing the enemy, cutting Chinese supply lines, fighting at night, taking advantage of his familiarity with the terrain, and other guerrilla tactics—shades of strategies to be used by the Vietnamese in the wars of the twentieth century—Le Loi was able to wear down his opponents' larger forces. The coup de grâce occurred when he was finally able to defeat the Chinese in a set-piece battle, using Vietnam's war elephants against the horses of the Chinese army. Le Loi then went on to become emperor of Vietnam.

One day, while taking a boat trip on the lake, Le Loi encountered a seven-foot turtle—a sort of demigod—which seized the magic sword in its mouth and returned the sword to its rightful place in the depths, now that the crisis was over and the country was independent again. Ever since, this cloudy green body of water has been called Hoan Kiem, or "Lake of the Returned Sword."

Enacted on a watery stage, the whole story seems part of a charming myth, like the legend of King Arthur and his sword Excalibur. Or, on a different level, like that of George Washington and the cherry tree. Like many myths, it is part morality tale, used in this case to demonstrate to schoolchildren the spirit of patriotism and exemplary behavior, or *chinh nghia,* meaning "just cause." Le Loi is a major nationalist hero, against whom all leaders are measured (and usually found wanting).

To an outsider, while there is no doubt about the historic Le Loi and his prowess as a general, it is nigh-impossible to conceive of any of the turtle-and-sword part being true, especially after a look at that lake. I don't know about a magic sword, but I found it hard to believe that anything—especially creatures like huge, century-old turtles—could live in this green, slimy, muck-filled, heavily polluted body of water in the center of a major city, just steps away from the motorcycles and pedicabs, alias *cyclos,* on the main drag. Other Westerners whom I relate the legend to have the same immediate reaction.

Except that researchers have found that the lake contains at least one giant turtle. And not just any big turtle but a special one indeed, involving one of the more astonishing success stories related to the EPRC, with as many cliff-hanging twists as any soap opera. It involved rescue efforts with an animal that was at the even thinner edge of extinction than langurs, and probably one of the most unlikely successes anywhere in Indochina. And it took place not in the jungle, but in the middle of one of the country's largest cities, surrounded by traffic.

On a typical day, you can see large numbers of people milling alongside the lake, which acts like Hanoi's version of New York City's Central Park. In the early morning, locals practice the Vietnamese form of t'ai chi in unison while joggers sprint by. Later in the day, passersby purchase ice cream at upscale eateries such as Fanny's on the west side of the lake, while on the opposite shore, tourists line up at the giant Soviet-era concrete monstrosity of a post office to buy stamps. On the north side of the lake is a graceful arched bridge to a small island, upon which sits a pagoda-style museum; the structure contains a stuffed turtle said to have weighed more than four hundred pounds when it was caught here in the 1960s. To the south is another, smaller island with a three-story building called the Turtle Tower on it—probably the most-photographed symbol of Hanoi. Just past the southern border of the lake is the main east-west street, Hang Khay, which leads eastward to such prestigious attractions as the Hanoi Opera House and the French colonial–era Metropole Hotel. (The hotel was recently restored to the height of its architectural glory, looking much the way it did when Graham Greene stayed there and wrote *The Quiet American*.)

All activity on the lakeshore stops, however, when someone spots what could be the nose of a turtle sticking up out of the water. A crowd rapidly gathers on the water's edge, with everyone craning their neck to look.

Legend has it that seeing this particular turtle in this particular lake brings good luck for a year, said Bui Dang Phong, vice director of Cuc Phuong National Park and the head of its turtle rescue program. Trained in the United States under a Ford Foundation grant, Phong attended Worcester Polytechnic Institute in Massachusetts for a PhD in international development, and is one of a new breed. As he puts it, "More Vietnamese biologists should be doing wildlife protection for the country of Vietnam."

"Traditionally, turtles are one of the four sacred animals in Vietnam, along with the dragon, the phoenix, and a half lion/half dog creature," Phong told me in an interview. Of them all, the turtle is probably the most esteemed, not only for its longevity but for its reputation as a benevolent protector of the kingdom of Vietnam.

Giant turtles carved out of stone embellish the thousand-year-old Temple of Literature, located at the other end of Hang Khay from the Opera House. Dedicated to Confucius in the early 1100s, the temple honors the country's finest scholars, poets, historians, and writers; atop each stone tortoise is a tall stone stele chiseled with the names of each individual and the date they earned their degree. In the 1400s, when someone passed their doctoral exam at the university, the emperor gave the newly accredited scholar a royal escort into their home village, complete with a band, horses, and elephants. Having their name carved into the list atop the stone turtles' backs at this temple was a key part of the honors.

Consequently, when a possible descendant of the legendary turtle of the lake is sighted, spectators armed with cameras jostle for position along the shore. According to at least one biological reference source—which calls the Hoan Kiem turtle a specimen of "the world's oldest and rarest softshell turtle"—the Hoan Kiem turtle is only seen an average of four times per year, making a sighting a difficult task. It has been suspected that one reason for the rareness of the sightings is that this species of turtle can remain underwater for long periods of time, by absorbing oxygen through its skin and the membranes of its throat. When it does come to the surface for air, it only needs to extend the very tiny tip of its nose above the water, periscope-like, to get oxygen.

Often, a sighting of this fabled individual turns out to be just a log or a piece of trash, but enough people have seen it for real to keep trying. And unlike the Loch Ness Monster, there is that actual stuffed version of the turtle a short distance away, along with bona fide sightings, authenticated pictures, and videotape. "Since 1991 the turtle has come up about four hundred times," claimed Vietnam's self-described authority on this individual turtle, professor Ha Dinh Duc (aka the turtle professor) of the Hanoi University of Science.

After years of speculation about whether or not the turtle really existed, the question was definitively laid to rest in April 2011, when the creature was finally captured alive for the first time, in a procedure involving dozens of people and two nets. Thousands of onlookers crammed the shores around the lake for a glimpse of the historic occasion, whooping and clapping when the turtle was captured; the crowd had to be physically restrained by police when it was taken to the island. A guard was posted near the site, allowing entry only to those with official accreditation, while researchers examined it in detail. (During one of its infrequent sightings, the animal was noted to have lesions on its shell and on its head, which prompted the effort to capture it. Researchers could also use this unique opportunity to determine its gender and its age, still unknown as of this writing. The best guess is that it is about one hundred years old, although some romantics like to say it is the original animal from the time of Le Loi, which would make it well over six hundred years old.)

Duc considers it a brand-new species, which he named *Rafetus leloii.* Other biologists say it is more likely a male of the species *Rafetus swinhoei,* a large freshwater softshell turtle. It is about five feet long and weighs about 220 pounds, with a leathery exterior, piglike snout, and flattish dorsal shell. Both its upper and lower shells, or "carapace" and "plastron," are covered with a layer of thick, leathery skin, hence the name "softshell." A strong swimmer, this type of turtle typically hides in the silt at the bottom of bodies of water, striking fish, frogs, mollusks, and other prey with sudden ambush attacks.

Regardless of the scientific name, the turtle in Hoan Kiem Lake was long thought to be the only one of its kind left in the world, said Tim McCormack of the Asian Turtle Program, an organization affiliated with the Cleveland Metropark Zoo that works closely with Bui Dang Phong. Born in the United Kingdom, with a zoology degree from Leeds University and a specialization in telemetry, McCormack has been working in Vietnam for eight years. When asked why he uprooted his life in Great Britain to rescue endangered animals here, he said, "It's much more exotic here, and you can do so much more."

How this one *Rafetus swinhoei* managed to survive here is hard to understand, as Hoan Kiem Lake is less than a half mile long, two hundred yards wide, and only six feet deep, with most of its shoreline covered with a concrete embankment. There's no sand in which to lay eggs, and no vegetation in which to hide. Finding a giant turtle here is like discovering a unicorn in New York City's Central Park—a single, unattached unicorn, with no prospect of a mate.

It seemed amazing, if sad. With no known females of its species, this animal seemed destined to mirror the fate of "Lonesome George," the Galapagos Island Tortoise who was the last of his subspecies and died recently. Besides the Hoan Kiem turtle, there were only two other turtles of this species, both male. One of the males lived in Dong Mo Lake, east of Hanoi, and the other lived in a zoo in China's Jiangsu province.

Then, in 2008, something miraculous happened. Biologists found an eighty-year-old female *Rafetus swinhoei* at another Chinese zoo, located in Hunan, China. Nicknamed "China Girl," it had been purchased from a traveling circus a half century earlier and remained unidentified until someone noticed how much China Girl resembled the picture in a notice on a Wildlife Conservation Society "Urgent" circular.

What's more, while no spring chicken, she was still able to lay eggs.

"You can imagine the excitement," said McCormack.

The male turtle from Dong Mo Lake was moved to the Hunan zoo in China and slowly introduced to China Girl. The zoo's pool was divided into three sections, with China Girl at one end, the Dong Mo Lake turtle at the other, and an empty section in between. On the second day, he was moved into the middle, and the two began sniffing each other through the fence. Considering that the creatures had not seen another turtle of the same species in about fifty years, things went well, and mounting attempts started in a few days.

She has since laid three hundred eggs—still unfertilized as of the time of this writing, but biologists were pleased with the progress.

McCormack says the lesson to take away is that even the most unlikely places still harbor something, and it is too soon to give up the ship.

A salutary side effect from all the attention was that the Hoan Kiem turtle's home waters are now starting to be cleaned up, through the efforts of a massive number of volunteers, experts from the Dresden Technology University of Germany, and scientists from the Hanoi-based National University. The group fabricated a custom-made floating barge that is slowly vacuuming up mud from small sections of Hoan Kiem Lake at a time, cleaning it, and filtering the water over a three-year period. Below the barge a "subaquatic vacuum cleaner" crawls along the lake floor using two corkscrew-like spirals that dig up and funnel the mud into a pipe, while also moving the whole apparatus. The goal is for a low-impact environmental technology that is silent and causes minimal turbulence while gradually improving the lake and avoiding the release of any toxic compounds.

While pleased, McCormack at the same time feels that the situation with the *Rafetus swinhoei* turtles shows how dangerously close to extinction the animals are. It also shows what last, desperate measures are needed: "When a fisherman found that 150-pound male in Dong Mo Lake outside Hanoi, it took hours and hours of patient negotiation to get it released. Here it was, one of only four in the world, and the fisherman didn't want to give it up. He finally did, telling us, 'It's the right thing to do—but I still wish I had the money.'"

Despite being a sacred animal—possibly why the Dong Mo fisherman donated it, and why turtles were traditionally unpopular as a foodstuff—turtles as a whole came under siege in the 1990s, when the country began to really open up to the outside world and trade with a booming China began to surge. The commerce in turtles built rapidly, and the non-profit Education for Nature Vietnam (ENV) recorded 434 cases of illegal hunting, trading, and smuggling of turtles since 2005. The numbers are so large that officials could not even count the turtles as individuals but instead record them by the ton; more than twenty-five thousand tons of turtles were seized by authorities from 2005 to 2010.

The reason for the trade is simple: money. The animals are popular for some forms of traditional Chinese medicine. Phong notes that one kilogram of the Chinese three-striped box turtle, or *Cuora trifasciata,* is reputed—wrongly—to cure cancer, and worth $5,000 on the open market. By comparison, that means it is worth more per kilo than heroin or cocaine, and with none of the harsh legal repercussions.

Other, less exotic, turtles are destined for the cooking pot as Chinese gourmet food. "You can't really blame the local people," says Phong; "they can get paid as much as a month's wages for a few turtles of some species."

Consequently, the Vietnamese pond turtle, *Mauremys annamensis,* has seen its population decline dramatically. Lately, rangers have seen a decline in the trade. "Not because we do better protection, but because there's fewer turtles around," opines Phong. Ten million turtles of all species are traded in Indochina annually, he says, with most going to China and some going to the West for hobbyists and collectors.

To combat the problem, a brand-new Turtle Conservation Center was established in March 2010 at Cuc Phuong National Park, under Phong's care. Located just a few hundred yards away from Nadler's EPRC and using his work as a sort of rough template, it is what Phong characterizes as "turtle heaven." Here, the staff takes care of turtles that have been confiscated, breed new ones, and attempt to educate the public about turtle protection. The center conducts radio-tracking programs to learn more about the basic biology of different species, and maintains a healthy population of 1,200 individuals of twenty different species.

Here, in an environment free from traps or hunting dogs, the population of each species can start to recover, for possible eventual return to the wild. In ponds full of clean, fresh water, without pesticides and chemicals, but with incubation rooms, terrariums, and plenty of water hyacinth for food and cover, the animals thrive, all on an operating budget of less than $12,000 per year. Most of the aid and technical advice comes from the Vietnamese government and from non-profits such as Fauna and Flora International, and the EPRC. Help also comes from the help of Westerners such as McCormack and Doug Hendrie, technical adviser to the ENV.

A self-guided outdoor nature trail runs through the Turtle Center, where visitors can view forty different turtles in 120 square yards of their natural habitat, surrounded by small ponds, running water, leafy vegetation, and undergrowth. The fences dividing each outdoor pen are artfully concealed; next to each of them are informational plaques, along with nonfunctioning examples of some of the traps that turtles must avoid: log-fall traps, wire snares, spring traps, hoop traps . . . the list is endless. It's an impressive display, and mindful of what I had learned about the lore of turtles in Vietnam, I found myself donating a small amount of money and purchasing two small jade turtles at their gift shop as souvenirs.

Probably the most important thing the center does is education. On a recent visit, schoolchildren could play a game that tested their awareness. Throw the dice, land on a square, pick a card, and try to answer the question, such as "What is the turtle's greatest enemy in Vietnam?"

Answer: "Human beings."

But children are not the only ones learning about turtles; the center also conducts training programs for rangers, police, and customs officials. Phong explains that the need for such programs became apparent in 1998, after a particularly big illegal shipment of land-dwelling tortoises was seized by customs. Not knowing what else to do, the officials simply dumped them into the nearest river. Unfortunately, while turtles can swim, tortoises cannot; consequently, all of them drowned.

"A little information helps," Phong said wryly.

Accordingly, the center conducts weeklong seminars to help customs officials and others distinguish among the different species. During the time I was at the park, a group of about two dozen students had completed the intensive turtle-identification course that McCormack was teaching, formally known as the "Sixth Annual Tortoise and Freshwater Turtle Field Skill Training Course."

To celebrate their graduation later that evening, these future authorities held a celebratory dinner, during which they gave a toast that they had composed especially for the occasion:

Softshell turtle
Softshell turtle
Hardshell turtle!

The evening then progressed to sessions with a karaoke machine that was hooked up to a powerful loudspeaker system. For some reason, John Denver's "Country Road" was a karaoke favorite that night, as it is throughout Vietnam. There is nothing stranger than hearing lyrics about rural West Virginia blasting through an open door onto a market square in a small city on the Vietnam/China border while a traditionally dressed ethnic Red Yao tribeswoman tries to sell you an unsolicited package of raw opium grown on the hills outside town. Visions of a Vietnamese version of *Midnight Express* dancing through my head, I hurriedly departed, sans extract of *Papaver somniferum*. (The local ethnic minorities have traditionally grown poppies in these hills for hundreds—if not thousands—of years.)

Over the noise of the singing, and in accordance with tradition, each student then proceeded to have a ceremonial, one-on-one shot of potent Vietnamese brandy with their professor, Tim McCormack. In line with the country's Confucian heritage, teachers are held in very high esteem, and this small gesture of a celebratory drink counts for a lot.

As McCormack and others explained to me later, it is considered rude for the teacher to refuse, and it is not sufficient for the professor to simply raise a single glass and toast the entire class. Instead, the teacher must drink a separate shot with each of the twenty-four individual students. It's not too demanding from the students' point of view, as each student just has a single drink with their instructor. But from the teacher's perspective, that amounts to two dozen drinks of very potent, high-octane alcohol in a row, in a very short period of time.

McCormack says he considers it a small price to pay for turtle conservation, even though he has to repeat the entire ceremony every time the Turtle Center gives a class.

Nevertheless, as yet another student beckoned, he groaned at the thought. Turning to me, he said, "When I die, I want my gravestone to say that I gave my liver for science."

Note: In a May 2011 press release, officials said that the Hoan Kiem turtle's wounds had healed and that it would be returned to its namesake lake. They also said that they had discovered that the turtle is a female, not a male as previously thought. If true, that means that there are now two known females and two known males of this exceedingly rare species—an astounding turn of events that lends credence to the turtles' reputation for good luck and divine intervention. It remains to be seen whether the turtle of Hoan Kiem Lake can lay fertile eggs as of this time. Its age is yet to be determined.

15

Why the Rhino Went
Extinct in Vietnam

Understanding what a Vietnamese really means when he answers
"yes" takes a few years of practice and experience.

—*Dos and Don'ts in Vietnam*

IT WAS THRILLING to see how Tilo Nadler's efforts with langurs had succeeded, and to see how rescue efforts at the very
extreme—when you're down to only four individuals of a turtle
species, you really can't go much lower—were progressing. In the
South of Vietnam, endangered cranes were making a comeback,
gibbons were being rescued, and all manner of Indochina species,
from butterflies to giant fish, were the subjects of study and rescue.

Meanwhile, in other parts of the world, some plants had been
brought back from the brink: in one case, there had been only *one*
known specimen of a kanaloa plant (*Kanaloa kahoolawensis*), which
had been brought back to a viable population. The sole member had
been found growing in the wild on a small Hawaiian island that
had been used as a bombing range for more than half a century.

Bringing things back from the near-dead has happened often enough that some call it the Lazarus Effect.

Considering all these successful projects, it was initially a puzzle to me as to why the Javan rhino, the symbol of and the reason for Cat Tien National Park, should have become extinct in Vietnam.

But the more I discovered about the difficulties and challenges of preservation, the more it made sense, especially after I learned more about the characteristics of rhinos and reread some old Environmental Science 101 textbooks. It all came down to a matter of basic ABCs, or what biologists refer to as "K" and "r." Rhinos are not fast breeding like rabbits, whose numbers can swell seemingly overnight, with little effort expended by rabbit parents to raise their young. These fast breeders—which include turtles, fish, frogs, and insects—depend on producing vast numbers of young, in hopes that at least a few will survive, in what is technically known as an "r-selected strategy." For example, about two dozen rabbits were imported from Europe to Australia in the 1850s by a hunting club in order to provide better sport for its members. The land down under had never had rabbits previously, but the animals and their r-selected strategy worked tremendously well, to the point where there are now an estimated 200 million wild rabbits in Australia, crowding out the native species and denuding the landscape. The rabbits' strategy—to overwhelm any potential predators and out-compete their rivals through sheer numbers—certainly succeeded, even if we wish they had not done so well. Efforts to bring down their population and restore Australia's environment to its previous state after all this time has passed are difficult and contentious, as I observed while covering the issue as Australia correspondent for the journal *Science*.

In contrast, rhinos and other large mammals rely on what is known as a "K-selected strategy," in which they produce only one or a few offspring at a time, concentrating on investing in the rearing and protecting of their young in hopes that while they have fewer offspring, more of them will survive to become parents themselves. A rhino is putting all its eggs into one basket—and watching that basket very closely, doing all it can to ensure that what few young it has will make it.

In a sense, it's a question of quantity of young (r-selected) versus quality of upbringing (K-selected). Both strategies work, but there is a key difference: when a population of rabbits gets down to low figures, they can rapidly bounce back. But when K-selected species drop to these low figures, they're in trouble, particularly if it is a species of large individuals that bear very few young and have extraordinarily long gestation periods.

When does a K-selected species, such as the Javan rhino, cross the point of no return?

Writing in a popular Australian science magazine called *Cosmos,* Mike Gilpin, a population biologist at the University of California–San Diego, explained that it is tricky to determine the absolute smallest number of a species needed to produce a viable gene pool to keep it functioning over the long term. A species needs a variety of individuals in order to avoid a genetic bottleneck. Using *Homo sapiens* as an example, Gilpin wrote:

> You need a minimum of 10 to 50 individuals to ensure against chance events and thus maintain a population over a relatively long lifetime. But this number assumes a constant environment, which is never true in nature. Environmental variables can be buffered only with much larger populations, sometimes orders of magnitude larger. In addition, you want to subdivide the species into core populations in several different areas, if only because of the threat of disease.
>
> Any small population is also subject to genetic risks, the gradual accumulation of traits with small but deleterious effects. Having about 1,000 individuals would secure against this eventuality for thousands of generations. In a self-sustaining

technological paradise, this number would probably keep us going until the next major asteroid hits. All bets would be off, however, if we fell back to a culture of primitive agriculture or hunting and gathering, as environmental vicissitudes would then present far higher extinction risks.

As if this was not enough to stack the deck against the rhino, their reserve in Cat Tien was located rather close to the largest city in Vietnam, so there was no possibility of the rhinos taking advantage of any of the "border effect" as described earlier.

Other factors were items that Gert Polet alluded to back in the '90s. I had expected him to make a pitch for building up the infrastructure and services needed for mass tourism: bridges, roads, sewage systems, places to stay and food to eat, the kinds of things Cat Tien noticeably lacked at the time. Instead, Polet said, he hoped that any physical improvements would be delayed or at least screened carefully to ensure that nothing was done to upset the local ecosystem. Polet saw Cat Tien's isolation at the time as a welcome natural barrier that protected its fauna and flora from destruction. "If you build a road or bridge, you'll just bring in more cars and motorbikes and put more pressure on the park," he noted. In 1998, the park had an average of only one Westerner per day, and only three or four Vietnamese tourists. Since then, like many such reserves in Vietnam, local officials have learned that tourism can be a gold mine, and in the country's overall pell-mell rush to develop, a lot has been invested in infrastructure that caters to tourists. The park now gets more than eight thousand visitors annually, and has recorded 25 percent yearly growth in the past few years.

The emphasis is on funding buildings, bridges, and other physical structures instead of things like hiring and training staff, setting up regular patrols, running interpretive programs, developing restoration projects, or other conservation efforts.

In addition to catering to tourists, infrastructure improvements are easy to measure. After all, it's very clear when a new bridge has gone up, and government agencies know what they've gotten for their money. In contrast, it's harder and more subjective to measure how much more biodiversity is in the landscape, or how much local people have embraced the idea of responsible, genuinely sustainable development. "We're building guardhouses but we don't have the guards to sit in them. We have motorboats to patrol the river but we don't have the people to drive them . . . I am talking much more about people, while the government is talking much more about a fence," Polet had told me then.

These trends, and the accompanying loss of habitat, are things that Vietnam's animal wealth can least afford. Vietnam's population has boomed, from an estimated thirteen million people in 1901 to about ninety million as of today, according to the Asian Historical Statistics Project. Once the country was mostly forest; now much of it has been cleared to provide timber and cropland. Nearly 56 percent of the country's forests have been leveled in the past fifty years. The remedy may lie in small-scale, sustainable harvesting of the trees, the replanting of native plant species, and public education.

But meanwhile, the local villagers are eager to exploit the land's resources and jump into the modern era, at the expense of protecting their natural environment. Polet was somewhat sympathetic to their plight. "They want the things that money can buy," he said. "A moped, a house that does not leak."

But consumerism is not the only thing driving local people to overexploit their animal resources. Polet argued that Asian society sees a different relation to nature than the one common in the West. Many Vietnamese seem to regard national parks as primarily meant for noisy kinds of recreation, rather than conservation. "They see it as an amusement place," Polet said. "They expect to see bumper cars here." He went on to say that the general Vietnamese public liked the idea of a park that was commercialized, tamed, and under control. "Otherwise," he said, "it is just like what they escaped." The benefits of a noncommercial, wild, and untamed environment—vital to endangered species—are not immediately apparent.

In addition, ecologists in Vietnam face some of the same issues that plagued the US Army in Vietnam decades ago: an overreliance on physical structures, concerns about the forced relocation of villagers to places outside the park boundaries, charges of nepotism, and cultural clashes between East and West. These challenges are well known to field researchers such as Polet, but the difference between his situation and others is that he had so few rhinos to begin with, and these animals are one of the ultimate K-selected species, little able to bounce back from a low population. And even the best-intentioned national policies cannot work if local people do not respect the regulations handed down from the capital.

"The decrees of the emperor end at the village gate," was the same woeful refrain I kept hearing throughout Vietnam.

In contrast, Tilo Nadler had several advantages: the EPRC may have been a new project, but it was in Vietnam's oldest and most established national park—founded by the George Washington of Vietnam, no less. Meanwhile, Cat Tien had only been a national park since the late 1990s, putting it thirty years behind Cuc Phuong in terms of setting up protection programs and patrols; doing wildlife surveys; creating rescue centers, veterinary clinics, and education centers; establishing relations with the scientific community overseas; bringing in researchers; and building local support. And for the purposes of large, wide-ranging animals like rhinos, the park was still considered small by international standards.

In addition, Nadler has been working with langurs for more than twenty years, while Polet was only given a three-year contract to work on rhino conservation. Geography also helped: Cuc Phuong was farther away from any urban area, so it was relatively isolated from the pressures of mass tourism. What's more, Cuc Phuong had never been bombed during the war, unlike Cat Tien, many of whose native plants had been wiped out by the effects of na-palm and Agent Orange. With the native brush and undergrowth replaced by less nutritious or suitable bamboo, rattan, Australian eucalypts, and American grass, the wildlife suffered. Furthermore, Nadler had been able to build strong community ties—even mar-rying the daughter of a local government figure, who in turn had some pull with the local Party authorities. These were advantages that Cat Tien simply did not have.

Merely getting a good handle on the number of rhinos was problematic. Despite their enormous size, the animals can easily disappear in the dense second-growth rattan brush. A lot of the bio-logical fieldwork was iffy, quick-and-dirty surveys of what people claimed they saw. Polet illustrated this with an example:

POLET: *Did you see a tiger?*
INFORMANT: *Yes.*
POLET (showing a photo of a leopard): *Is that it?*
INFORMANT: *Yes.*

Yet whenever he suggested that the number of rhinos was fewer than previously thought, the attitude of many in the scientific com-munity was that there simply had to be more hiding in the depths of the park. Nicholas Wilkinson of the Darwin Initiative Project in Hanoi is familiar with the situation: "The problem is that if you don't find something, then people simply say you didn't look hard enough, or looked in the wrong place. This can go on endlessly . . . and it raises the question of 'How do you prove that something does not exist?' How do you prove a negative? Sometimes, I feel a lot like a doctor in the sixteenth century: we have a lot of ideas but we don't really know anything."

Combine all these factors with a smaller number of animals to start with—in the case of Delacour's langurs, one of the more endangered primates, there were two hundred animals as opposed to possibly as few as five or seven rhinos at the start of the rescue program—and the playing field begins to look seriously tilted against the rhinos' survival. The fact that langurs can bear young in only seven months as opposed to the sixteen months of a rhino further tilted the odds. Add in the fact that the rhinos' horns were so tempting to poachers, and the conclusion is foregone. The table, constantly tilted, could tip no more, so it fell and collapsed.

As the extinction of the last Javan rhino was only confirmed in late 2011, it is too soon to say what the repercussions will be with the loss of Cat Tien's signature species. Hopefully, now that the park has been in place for several years, it will survive. The loss of the Javan rhino may provide a warning, and spark greater efforts to restrain the black market and get the park's problems under control. Otherwise, the rhinos' demise will be just another needless extinction on the Asian mainland, of which there have been far too many.

And who knows what else became extinct before we were even aware of its very existence? At the current rate of extinction, species are disappearing worldwide before they can even be found or named. There are hints and rumors of many interesting creatures still hiding in Vietnam's Lost World, some of which may be fiction—or may be fact.

16

Vietnam's Yeti:
What Else Is Out There?

One of Kien's MIA team said he had heard crazed laughter echoing
from Hill 300, on the other side of the Sa Thay River. Kien listened
as the nervous man gave his version.

"I think it came from the jungle monster the Trieng people
talk about," said the soldier. "Anyway, I'm sure it wasn't a
human laugh, because it was shaking and choking.
It didn't last long but I froze in my tracks . . ."

"Perhaps it was the Forest Man," said another,
remembering the local folklore.

—Bao Ninh, *The Sorrow of War*

THE EXTRACT IN the epigraph to this chapter is from a semi-
autobiographical novel written by a former infantryman who had
fought on the opposing side during what we Americans refer to as
the Vietnam War. Of the five hundred draftees of Hanoi's "Glorious

27th Youth Brigade" who went south to fight in the Central High-lands, the author, Bao Ninh, is one of only ten who survived to return home. Afterward, he became a member of a team assigned to recover the corpses of soldiers, and during that mission they had this possible near-encounter with the Forest Man, or Nguoi Rung.

The Sorrow of War can be found everywhere, including souvenir shops, newsstands, backpacker hostels, Westerner-oriented restaurants, and anywhere else foreign tourists can be found. His slim paperback is usually sold paired with another book of similar size and page-length but with a different viewpoint: *The Quiet American,* by Graham Greene.

Officially banned for many years, the copies of *The Sorrow of War* that you see are often cheap photocopies of photocopies, poorly bound together with glue, much like old samizdat literature from the Soviet Union during the Cold War. The reclusive author is one of the few voices to speak candidly from the perspective of the Communist side of the human cost of the war; it's a miracle he was not jailed.

The Forest Man he referred to is a translation of Nguoi Rung, sometimes spelled "Nguoi Rung." That it should appear in the most popular war novel ever written in Vietnam shows how powerful a hold the creature has on the imagination. Even the name Forest Man is tantalizing: in Indonesia, "orangutan" translates into "forest man" in the local people's language.

Reports of Nguoi Rung sightings, along with its supposed foot-prints and casts, have occurred often enough that authorities have gone looking for the creature. A French colonist recorded spot-ting *l'homme sauvage* in 1947, and there were so many reports of the creature in the Parrot's Beak section of the Central Highlands in 1974—the region where Vietnam, Cambodia, and Laos meet—that Hanoi officials authorized an expedition to the region by the most prestigious biologists in the country, even while the Second Indo-china War raged.

But the majority of zoologists are taking a wait-and-see attitude about Nguoi Rung. Meanwhile, among the public, the scientific tussle has only added to the animal's attraction, and added yet another element to stir the pot.

Asked about the possibility of the creature's existence, Ben Rawson of Conservation International said, "On the face of it, it's easy to say it's absolute rubbish. It falls into the place where science and folklore converge. Some of the stories are truly strange, such as that the Nguoi Rung has short arms so that it can kill people better, it can mimic any call, its feet are on backwards, and it cannot get up if pushed over."

At the same time, some think there may be some kernel of truth behind it all, once the folklore is stripped away. Rawson admitted, "It could be some terrestrial mammal, something like a type of orangutan. People like Le Vu Khoi have been looking for it, and Vu Ngoc Thanh, a curator at Hanoi University, has casts of the footprints of some kind of creature." (Thanh did not respond to any subsequent attempts to reach him.)

When I asked another researcher, Nicholas Wilkinson, he replied that scientists can find the very effort of looking for mythological species to be irritating when there are so many bona fide, known, rare species, such as the kouprey. He sees no problem with stumbling across new species while doing broad-based, region-wide general surveys, but sees targeted quests for the Nguoi Rung as ill-considered: "Personally, I think it is not the best way to spend one's time and resources. Still, I understand the mystique."

Meanwhile, reports continue to pop up, both old and new. In South Vietnam, during the war years, at least one US soldier reported firing upon a Yeti-like creature at night in the Central Highlands, not far from the huge US military base at Da Nang, said a widely quoted old copy of the *Army Reporter* on the Internet—although at this late date, it is hard to know what happened. The creature might have actually been an enemy fighter hiding in the darkness, or the incident could well have been a case of a tired, scared young soldier firing at an innocuous object in the night.

Still, decades later, the stories persist. In 2002, Michael P. Kelley, author of *Where We Were in Vietnam: A Comprehensive Guide to the Firebases, Military Installations and Naval Vessels of the Vietnam War,* wrote, "This author's unit found what appeared to be a dead Orangutan on a landing zone on Nui Mo Tau, southwest of Hue in '70, and later we were told there were no Orangutans in South Vietnam."

Biologists have a term to describe the hunt for such legendary creatures, calling this field "cryptozoology"—literally, the study of hidden animals. The word was coined by Belgian zoologist Bernard Heuvelmans from the Greek word *kryptos,* meaning "hidden" or "unknown." Relying as it does upon thirdhand reports, legends, folktales, circumstantial evidence, and alleged sightings by amateurs, it can be hard to take seriously sometimes, as it lumps legitimate sightings by experienced observers together with sightings of Bigfoot and the Loch Ness Monster. It is not a recognized branch of the formal science of zoology; nevertheless, there was until a few years ago an International Society of Cryptozoology, with about nine hundred members, including some distinguished scientists.

More recently, in September 2011, a group calling itself the Center for Fortean Zoology (CFZ) announced that its founder, Richard Freeman, was going to Sumatra to write up a series of newspaper articles for *The Guardian* on his effort to pursue reports of a four- to five-foot-tall apelike creature that walks upright, known to locals as the "orang pedak." The UK-based group's website says that it is "the largest professional, scientific and full-time organisation in the world dedicated to cryptozoology"; it also sponsors a "Weird Weekend" devoted to the latest sightings; and it is based "in an old country house in rural Devonshire, parts of which are well over 200 years old. It is home to several ghosts, a bewildering and ever changing array of animals . . ."

It seems that few people want to categorically rule out any new "hidden" creature, given the number of animals that Western science did not know about until very recently.

After all, the saola was identified in 1992, and the giant muntjac in 1994. Between 1997 and 2007, at least 1,068 new species have been discovered in the Greater Mekong, two new species a week on average every year for the past ten years.

The story gets even more convoluted when dealing with creatures that we definitely know had once existed but were only recently exterminated. The kouprey was alive and well in a Paris zoo until 1937, and there is film footage of it in the wild from the 1950s, photos from the 1960s, and reported sightings in the '80s and '90s. No one can definitively say whether it is extinct today.

The phenomenon is not confined to just Indochina. In the United States, the world of ornithology was electrified in 2006 when there were credible sightings of the ivory-billed woodpecker, or *Campephilus principalis*—a bird previously thought to be wiped out—in an Arkansas swamp. It started a comprehensive, systematic five-year search by a six-person mobile search team from Cornell University's Laboratory of Ornithology. (The jury is still out on the ivory-billed woodpecker's continued existence.)

And in the Australian island-state of Tasmania, the Tasmanian tiger, or *Thylacinus cynocephalus* (Latin for "pouched dog with a wolf's head"), was alive and well in a Hobart zoo as late as the 1930s. Rumors persist that the species is thriving in the dense undergrowth in the wilds of the island's west.

No irrefutable photographs, fur, or plaster casts of tracks of the Tasmanian tiger have provided confirmation, but such tantalizing sightings have helped to make the animal, sometimes simply called the Thylacine, into a Tasmanian obsession. When I was there to do a story for a science magazine, I found that images of the two-foot-high, shy nocturnal predator were everywhere, including city seals, traffic signs, T-shirts, and beer bottles. The Tasmanian

Parks Service receives notice of dozens of sightings every year; a ranger systematically tallies and evaluates all of them.

Part of the animal's mystique is the nature of its demise. The world's largest marsupial carnivore disappeared recently enough that there are still hunters around who remember killing it for the two-dollar bounty. (It was a voracious killer of sheep, the worst of all sins in Australia.) The Thylacine was not protected until two months before the last one died in captivity. "There's almost a guilty conscience about its disappearance," said Mark Holdsworth of the Parks Service. His colleague Steve Robertson agreed. "It's the idea of redemption. We killed it off, but now it's back."

For his part, Holdsworth finds large-scale searches for the tiger frustrating. He thinks the focus should instead be on protecting existing endangered species. There is only one benefit of the misplaced public interest in the Tasmanian tiger, Holdsworth maintained: "The Thylacine is a good reminder of extinction and endangerment. We're still making the same mistakes."

Meanwhile, a Tasmanian zoologist spent all afternoon giving me reasons why the overwhelming odds were that the creature had become extinct decades ago. But at the very end of our interview, he added that he himself had found what looked like Tasmanian tiger tracks just a few years earlier, in the wilds of the westernmost, most unpopulated part of Tasmania. . . .

What I took away from hearing about reports of sightings of the Forest Man in Vietnam or the Tasmanian tiger in Australia is that there is a romantic appeal to such creatures. Even the quixotic nature of these searches is endearing; it speaks to something in us.

In addition, it's hard to come down too harshly on these quests in a place where two new species have been found every week, fifty-two weeks per year, for the past ten years—with seven new large mammals found in Vietnam in a single decade, including the saola and the giant muntjac. At one point, they, too, were the subject of unconfirmed sightings.

Taken together, these factors make it hard to say what would be a pointless quest and what is worthy of a legitimate scientific survey. One person's rumor is another's solid lead, and there will always be quasi-reliable reports of new critters, confusing our understanding of what is out there in the mist-shrouded Lost World.

As long as the wild habitat is protected, there is always that potential for another new and electrifying wildlife discovery. But if the land is all paved over, that sense of mystery, and potential, is gone.

Two Futures:
Angkor Wat or Kauai?

Biodiversity is lost thread by thread. And then you lose your shirt.

—Chipper Wichman, National Tropical Botanical Garden,
Kauai, Hawaii

VIETNAM FACES SOME choices about its long-term future.

It could go the way of the ancient city of Angkor Wat next door, where the Khmer inhabitants of the capital of the old Cambodian empire overexploited their natural resources and saw their ecosystem collapse, leading to the demise of their civilization. Angkor had an exquisitely designed series of reservoirs and canals, aimed at hoarding the water from the monsoon season to guarantee a steady water supply in dry times, allowing the ancient Khmer kingdom to raise water-thirsty crops such as rice. In a word, Angkor was a "hydraulic city" entirely dependent on water; as many as 200,000 Khmer were needed to create the massive three-story dams needed to hold back the water from places such as the Siem Reap River.

But they expanded beyond the carrying capacity of the land, with no room to absorb an unusually dry year.

This pattern of overuse has shown up in numerous places. At Easter Island, the diminishing size of fishhooks over time speaks eloquently of a people forced to go farther and farther down the food chain in order to survive—small fish previously thought unworthy of catching became targeted once the population of bigger, more desirable fish crashed. Likewise, some theorize that the ancient Mayan kingdoms collapsed due to overexploiting their environment.

And modern-day Haiti has been called an "environmental disaster area." Even travel guides such as Lonely Planet's *Caribbean Islands*—hardly likely to downplay a tropical travel destination—call Haiti "a popular university case study in environmental degradation and disaster, perhaps equaled only by Madagascar and the Amazon rain forest. Unchecked clearing of the land for food production and fuel wood has depleted massive tracts of broadleaf forest . . . the destruction of the forests for firewood and farmland has caused an untenable amount of soil erosion, as well as trapping Haiti's peasants in a cycle of subsistence farming with ever-diminishing returns."

Such bare mountains can be dangerous to life and limb during the region's infamous monsoon season, when rain causes horrendous floods and mudslides. (In contrast, the next-door Dominican Republic, with its forest cover largely intact, comes out relatively unscathed.)

It's a dark scenario but not necessarily one that Vietnam is fated to follow.

What Tilo Nadler is doing at the EPRC with langurs, and what zoologists are doing with other species all over Indochina, could expand into something much bigger. Perhaps a good model for the long-term future is what the National Tropical Botanical Garden (NTBG) in Hawaii has been doing for nearly fifty years with endangered plants across the entire Pacific.

Appropriately enough, the NTBG is located on Kauai, known as the "Garden Island" of the Hawaiian archipelago, and fourth largest of the Hawaiian Islands. With a landmass about half the size of the state of Rhode Island, Kauai is home to lush, green, dramatic cliffs that rise more than a half mile above the ocean; a cliff-side hiking trail that is considered to be one of the ten best in the world; a large tropical reef; and a canyon—Waimea—to rival the Grand Canyon. About 90 percent of the island is still undeveloped; by law, no building can rise taller than four stories, about the height of a tall palm tree. The vibrant green scenery at one end of the island makes visitors understand why portions of *Jurassic Park* were filmed here.

The NTBG contains five gardens, three preserves, 1,800 acres, and 120 employees; all dedicated to aggressively finding and preserving endangered plants of the tropics. With just over half of its native plant species—more than 250 of its estimated 500—having become extinct since Europeans first made contact, the Hawaiian Islands are considered to be the "extinction capital of the world." (By comparison, a comprehensive report issued jointly in September 2010 by Kew Gardens, London's Natural History Museum, and the Missouri Botanical Gardens, found that worldwide, about one in five plants is considered at risk for extinction. Meanwhile, fewer than 10 percent of bird species are at risk, and about 25 percent of all amphibians. The International Union for the Conservation of Nature's "Red List" puts animals at about the same level of risk as plants.)

To try to reverse this downward trend, the NTBG spends a relatively modest $9 million per year, employing a number of strategies to save as many plant specimens as possible in as many different ways as possible: seed banks, pollen collections, dried plant pressings, DNA records, photographs, maps, GPS coordinates, descriptions in peer-reviewed journals, the cultivation of rare plants in greenhouses for eventual return to the wild, and the propagation of endangered plants that have been confiscated by US Customs. (Its Plant Rescue Center for seized black-market specimens has been quite busy; at the time I visited, the center had $100,000 worth of rare *Paphiopedilum* orchids in a single flat tray the size of a four-by-eight-foot sheet of plywood. Somewhat similar to the familiar "lady's slipper" of New England but native to the Pacific Islands and Southeast Asia, each individual plant was worth between $700 and $1,000.)

The NTBG sponsors ten-day workshops for science journalists in order to give them a sense of what the garden's staffers call "extreme botany." Participants live alongside field botanists and get the total fieldwork experience, down to the outdoor shower in the backyard, where wild pigs can be heard and seen rummaging in the neighboring bushes. Like all visitors to the islands, the journalists are greeted with a garland of leis—in my case, a handmade strand of purple-and-white *Vanda* orchids containing a faint but noticeable scent. Their purple/almost-blue color makes them especially prized by horticulturalists; as a result, many varieties of the *Vanda* genus of orchid are endangered because of overharvesting and habitat destruction. When I look them up online later at a wholesale florist website, I discover that I had $300 worth of flowers around my neck. Here, however, they are almost—*almost*—commonplace.

Many of the flowers here are found only in isolated nooks on cliff faces, where they are hidden away from the ravenous appetites of non-native goats, rats, and pigs. The sheer inaccessibility of their hideaways on the basalt cliffs is their only defense; the cliffs are too steep for predators. Many have no other natural defenses such as burrs, spikes, poisons, or nettles.

Because of rapid development elsewhere in the tropics, many of these plants have lost their habitat and face near-extinction. There are only twenty specimens of *Hibiscus delphus* known in the wild, and only four trees known of *Hibiscus clayi;* the only known wild specimen of the star violet (*Kadua haupuensis,* a member of the coffee family) disappeared down the throat of an intrepid goat.

Researchers know that at least 150 species consist of fewer than a hundred individuals in the wild.

Some of these plant species are highly specialized to take advantage of unique conditions found in each patch of "microclimate" on the island, much like the woodhens of Lord Howe. Such specialization in a favored niche means that a plant can outcompete its nearby rivals. But putting all its metaphorical eggs in one basket does make such specialized plants more vulnerable when a big generalist comes along; it's sort of the plant equivalent of what happens to a mom-and-pop specialty shop when a generic big-box discount store opens next door.

The penalties for those plants that have overinvested in the specialized conditions of a particular locale can be severe. Some plants, such as the iliau or *Wilkesia gymnoxiphium,* a relative of the silversword, take twenty-five to thirty years to go from seedling to adult; they blossom only once and then die. We see almost a hundred of these twelve-foot-tall, striking plants in bloom while hiking not far from the Awa'awapuhi Trail, along the dry ridges of the island's Waimea Canyon. Each plant consists of a two- or three-foot-high stack of dozens of smaller, yellow daisylike flowers on the end of a tall stalk, making it look like a more spectacular version of the head of a dandelion. When a large number of these plants bloom in unison, as they have here, it looks like a waterfall of flowers. One of our botanist guides, Steve Perlman, tells us that this is the best bloom he's seen in years. Disconcertingly, we also see evidence of what our botanist guides tell us are wallows of churned-up earth created by introduced feral pigs, which love to chew on plant roots. The ground is so torn up that it looks like it has been rototilled.

At first glance, it's tempting to say "So what?" Upon hearing of these conservation efforts, a provocative computer scientist friend who works at the birthplace of the World Wide Web, outside Geneva, Switzerland, asked, "Can't we just freeze their DNA and store it somewhere, for cloning later? Genomes can be rapidly and cheaply sequenced at a price less than a thousandth of the cost ten years ago, so we could just store those assemblages of letters on a computer disk. Couldn't all of life be reduced to nothing but a series of data points—collections of strings of the letters A, T, C, and G—adenine, thymine, cytosine, and guanine?"

It's a question that drives biologists crazy. When I pitched it to Ben Rawson of Conservation International, he replied, "First of all, things such as cloning, or freeze-drying DNA samples, are real last-ditch efforts. There's no real conservation benefit from it—you just have a bunch of genetically identical samples, not a healthy, robust population of separate individuals in a complete, living ecosystem. And you still don't know anything about how to keep it alive and producing young afterwards; you have no information about its reproductive cycle, dietary needs, social structure, or anything else."

He argued that wild animals are more than mere collections of genes. Each generation of animals teaches its young basic skills essential for survival in the wild, including such vitals as how to use tools, how to communicate with others, where and when to migrate, what to eat and what to avoid. A frozen genetic sample lacks any of this information vital for survival. It's like a computer without software.

Unfortunately, the mere presence of a genetic storage bank or "ark" somewhere can make people think they no longer need to care for the natural environment—if the animals' genetic materials are safely frozen, who cares about clear-cutting the jungle? But such projects do *not* mean that we have no need to worry about the disappearance of a species in the wild because there is a slight chance of resurrecting it later through cloning and cryogenics.

The staff at the NTBG—and projects such as the Svalbard Global Seed Vault, a massive cold-storage unit in Norway that contains millions of seed samples from around the globe—try to make it clear that such scenarios are last-ditch efforts, designed for doomsday situations, with strictly limited goals in mind. The same is true of the cryogenically preserved samples of sperm, eggs, embryos, tissues, and other cells that are kept in the "Frozen Zoo" at San Diego's Zoo; or the stored samples in the Millennium Seed Bank Project at Kew Gardens; or the entries in such informatics projects as the Encyclopedia of Life—the attempt to provide a centralized portal of data on the estimated tens of millions of species on Earth. These projects provide vital information for the scientific community, and they carry on the effort of Linnaeus to go forth and collect and organize, but they are in no way a replacement for conservation. Scientists such as E. O. Wilson of Harvard University say that there is still an opportunity to reverse some of the processes of man-made extinction and control the damage—determining whether we lose 10 percent of all species on the planet or 50 percent.

Losing any percentage does have real consequences: from a purely selfish, human-focused perspective, humans get many immediate, pragmatic benefits of wild organisms. Lose these species and we lose their many benefits. Plants and animals do have utility and value; you never know from what natural resource the next cancer cure will come from. A source in the pharmaceutical business told me that they were well aware that the best leads for new drugs come from nature's medicine cabinet: the heart medication digitalis was derived from a poison originally used on blow-gun darts in the Amazon; the active ingredients in aspirin are modeled after the compounds in willow bark, which Native Americans chewed to relieve headaches; and the extract from a plant known as the Madagascar periwinkle (genus *Catharanthus*) is used to treat the cancer known as Hodgkin's disease.

What's more, the more species of plants, animals, and other life-forms that live in a given region, the more resistant that region is to destruction, and the better it can perform its environmental roles of cleansing water, enriching the soil, maintaining stable climates, and generating the oxygen we breathe. Such things do have a dollar value; scientists such as Gretchen Daily of Stanford University have recently started seeking to quantify the value of this "natural capital" of nature's goods and services.

Because no one previously put a price on how a given piece of forest reserve provides flood protection, water for drinking and irrigation, hydropower production, biodiversity, carbon storage, or crop pollination, conventional economics tends to devalue it. In contrast, Daily hopes to show that there are indeed immediate economic benefits to keeping a forest intact. A forest is not worth money only when its trees are cut down; it is also monetarily valuable when left in peace.

Carrying the metaphor further, biologists such as George Schaller like to say that nature should be husbanded and nurtured, because it is in a sense society's "capital" in the bank. "We should be living wholly off the 'interest,'" he said. "People seem to be incapable of thinking long-term. Politicians only think ahead to the next election, and businessmen only think about short-term profit this quarter."

Mindful of Daily's work and thinking long-term, governments such as that of Costa Rica pay landowners to maintain native forest rather than cut it down. Known as the Payment for Environmental Services Program, it has helped to reduce the rate of deforestation in that country.

Meanwhile, to give the most endangered plants any hope of long-term survival, researchers must find the last few specimens and bring them back to NTBG greenhouses, where they can hopefully be bred in a secure, safe environment and their numbers increased. Once enough have been raised, the plants can be distributed across the tropics, over much of their original range, thereby lessening the chance that any one single catastrophic storm, natural disaster, or man-made blunder wipes them all out. Indeed, such a storm did strike the NTBG's six-hundred-acre Limahuli Preserve on the north end of the island a few years ago, destroying some of its rare native vegetation. (The storm scenes in *Jurassic Park* were shot at that time. Though that was nearly twenty years ago, the natural environment has still not fully recovered; many invasive species were quicker to colonize the storm-cleared openings in the forest.) To make sure something similar does not wreck the headquarters here on the south end, with its rare pressings and seed banks, the NTBG built a hurricane-proof, blast-proof, "green," naturally lit, climate-controlled storage building on its main campus, with its own backup emergency generators.

Restoring their seedlings to viable, adult populations may sound like an impossible task, but success stories exist. A single wild plant known as *Cyanea pinnatifida* on the island of Oahu was successfully bred at the NTBG and then its young "outplanted" to hundreds of healthy flowering plants; the original is now dead but its offspring live on, in carefully nurtured preserves.

In another case, a single specimen of a plant previously thought extinct was found clinging to life all by itself. In 1992, the NTBG's Steve Perlman and Ken Wood found the only living example of the kanaloa plant (*Kanaloa kahoolawensis*) growing in the wild, on Kaho'olawe, a small island formerly used as a bombing range by the US Navy for more than fifty years. When the island was no longer needed for live-fire training, the land was given to the State of Hawaii, where it is now a reserve. Before the handover, a $400 million cleanup campaign was conducted, which included the placement of used tires in gullies to stop erosion, the removal of unexploded ordnance and scrap metal by hand, goat eradication, and extensive archaeological and botanical surveys, during which a single specimen of kanaloa plant was found. The plant was previously known only from the fossil record.

As a result of such experiences of the Lazarus Effect, the NTBG takes a never-say-extinction approach to plant protection, no matter how hopeless the case may seem, or how seemingly insignificant the plant. "Biodiversity is lost thread by thread," explains NTBG director Chipper Wichman. "And then you lose your shirt."

And it's not just plants that have bounced back, and not just in Hawaii. Some remarkable recoveries from very low numbers have occurred, such as zoo breeding stock from a single pair, for example. Mammalogist/taxonomist Colin Groves explained, "The present zoo stock of Père David's deer was bred from an initial stock of eighteen; Przewalski horses from twelve; okapi from twenty-four; gaur from ten; Siberian tigers from thirty-nine (and in the first two cases, that's all we have—there are none in the wild). Each of these zoo stocks is now flourishing. There is even a mathematical formula to estimate the average amount of genetic variation needed in a founder population in order to sustain a species."

At the beginning of the twentieth century, the Indian rhino (*Rhinoceros unicornis*) had been hunted down to just twenty animals; today there are close to two thousand. Similarly, the American buffalo (*Bison bison*) had diminished to just a few dozen animals by the end of the nineteenth century; now it numbers about 350,000 thanks to captive breeding programs. Herds of American buffalo can now be found in some of the most unexpected places across the globe; a herd of about eighty lies just outside Geneva, making for a strange sight as they graze in the meadow with Mont Blanc, Europe's tallest mountain, in the distance behind them.

Such coordinated, wide-ranging rescue efforts sound simple and straightforward in theory, but can be difficult in practice. One researcher with a lot of experience in this exotic specialty is Steve Perlman, an extraordinarily fit professional collector for the NTBG, who cheerfully acquiesces to the term "extreme botanist." His career consists of dangling down the faces of cliffs to find the last surviving plant, sometimes with his wife watching from a helicopter. Fifty-nine at the time we met, with two daughters, he regularly rappels down sheer cliffs in the roughest terrain, spending twelve-hour days for months at a time away from home. ("Luckily," he said, "my wife likes plants.") In one typical summer, he spent six weeks in the Marquesas, followed by a month in Pohnpei, and then went off to Palau, New Caledonia, Rapa, and other places hard to find on any map.

"Our collectors stop at nothing but the end of their rope," says Chipper Wichman. Indeed, the lobby of the facility contains a six-foot-wide photograph of one of his colleagues hanging from the underside of a cliff, clinging to a rope with two hands while clenching an endangered flowering plant between his teeth.

For Steve Perlman, a normal workday consists of going to the top of a 1,500-foot cliff on the Na Pali Coast on the north of the island, tying a rope around whatever he can find—rock, tree, vine—and rappelling down. The goal is to keep his hands as free as possible to do botany, so he often employs a climbing shoe he found in Japan that has a generous gap between the big toe and the other toes; with the aid of this arrangement, his toes can more readily engage the rope. The rest of his gear is more conventional, consisting of a helmet, vest, water, and calcium to combat the ninety-four-degree heat. (He drinks three liters of water a day.)

His most useful tools recently, however, are old-fashioned 35mm film canisters full of various types of pollen and a small brush his wife got for him from a Mary Kay cosmetics catalog. Using them, he is able to hand-pollinate plants on-site—the latest approach in rescuing plants whose populations are down to the double digits.

The worst thing for him? "Bees, wasps, scorpions, and poisonous centipedes. They get in your hair or in your face when your hands and feet are busy, and you can't brush them away."

What else?

"I have to be more careful about what I tie the top end to. One time, after seesawing back and forth for several hours with the rope on the edge of a sharp rock, I climbed back up to find that the rope had sawn itself in half," said Perlman.

The rock climber in our group of journalists is appalled.

Nevertheless, despite his laissez-faire, unorthodox attitude toward climbing (Japanese one-toed climbing shoes?), Perlman has been very successful, and has several plants named after him, such as the *Silene perlmanii*.

Still, even his experience does not mean every trip pans out. "Sometimes, you'll go out to an area and go all the way down a rope and there's no plant. Everything is dead. And then, you hit this low. I've even said I think we need hospice training, because we're dealing with terminal patients and they die on you. If you see them for ten or twenty years, they're your friends, and you know what they look like. Then you come back and they're dead and dried up. I've gone back and actually witnessed extinction at least a dozen times. And then I think, yeah, I'm not coming here again. I'll go out and get drunk or something, because I've just lost a friend."

One reason why so much is being lost is that there was so much to begin with. Wichman says that 80 to 90 percent of all biodiversity can be found along the Tropic of Cancer or the Tropic of Capricorn, where there are warm, hospitable temperatures and decent rain year-round, with many climate regimes in a small geographic space and a year-round growing season. These areas have plenty of resources to exploit, along with booming economies and young populations.

To counter this, one strategy of the NTBG is to raise awareness of Hawaii's botanical heritage by promoting appreciation of the role of plants in traditional Hawaiian culture. They see the two things intertwined with the old, pre-contact ideals of *ahupua'a,* or "stewardship of the land."

"On a small island, you quickly learn to stop overharvesting resources, and properly act as custodians of the environment for the long term," claims Wichman. He sees a continuum between protecting the islands' natural heritage and reviving old-time Hawaiian skills such as using the islands' natural resources for canoe-making, traditional crafts, fishing, and agriculture. To him, it's not much of a stretch between such plant-based self-sufficiency and the traditional songs, dances, and language of these islands.

Accordingly, the garden brings in artisans such as a well-known local woman named Kumu Sabra Kauka, president and spokesperson of Nā Pali Coast 'Ohana, a grassroots non-profit organization dedicated to the preservation of the natural and cultural resources of the Na Pali Coast State Park on Kauai. Winner of an award from the Historic Hawaii Foundation, her organization tries to research and restore some of the old ways that can still be found on the more remote islands of the Hawaiian Archipelago; locally, it is involved in beach cleanups and archaeological preservation.

Dressed in a green muumuu, Kauka shows us how to make *tapa* cloth from the bark of the paper mulberry (*Broussonetia papyrifera*) bush—its soft green bark is easily peeled, like that of a willow in New England in the springtime. When pounded for several minutes, the strips of moist bark mash together like the fibers of paper pulp; when dried, they form a linen-like material. The same is done for the bark of the pandanus tree for making mats, baskets, thatch roofs, and canoe sails.

We also learn about traditional Hawaiian plant-based medicines, plant drinks such as *kava,* plants used in the terracing of the land to raise breadfruit, the creation of rock-lined pens for raising fish, and many other features of an old Hawaii that I never would have thought to look for as a tourist.

The last part of the program consists of a version of the once-banned *hula,* performed by children from the local elementary school, who wear flower leis and grass skirts over their T-shirts and jeans.

At the very end of the day, Kauka teaches us a few words of the old language. To help us remember them, she has us learn something she terms her "Mahalo" song. She says that the exact meaning is hard to pin down in English, but when I later track down an online dictionary to Old Hawaiian, it says that the original concept in its purest form ran along the lines of giving "thanks, gratitude, admiration, praise, esteem, regard, or respect." We all get out our digital recorders and cameras, prior to joining in.

She leads us all in song and then we start to put away our gear. Unexpectedly, Kauka continues on to another verse that she has not taught us. Luckily, the digital recorder is still on to catch it. In a clear, even voice, the sound drifts up above the trees to echo off the low-lying hills of this part of the Garden Island. Kauka lingers on the very last word.

"Mahalo," she sings alone, to the sky and to the Earth.

"Mahalo."

18

The Big Picture

Each species is a masterpiece, a creation assembled with extreme
care and genius. . . . Destroying rain forest for economic gain is like
burning a Renaissance painting to cook a meal.

—E. O. Wilson, Harvard University zoologist

THE RECENT MIRACULOUS discoveries in Indochina speak of
life's constant drive to persist, even under the harshest of conditions. Like the ailanthus of *A Tree Grows in Brooklyn* that manages
to force its way up through cracks in the pavement, the flora and
fauna of Vietnam survived against all odds in the most extreme
environments, including the environment of the battlefield. It's almost as if humanity has been granted a second chance to get the
ecosystem right.

"Everything we want, need, and use is dependent on nature,"
said George Schaller. If we destroy nature we destroy ourselves. As
a region is deforested, the number of species it can sustain shrinks
faster, and its ability to function declines as well. In some ways, an
ecosystem is like a pyramid: you can remove a certain number of

items from its base without it collapsing. But the more items removed, the more unstable the entire thing becomes, until you reach the point of no return and the whole system collapses and takes us with it. (Another analogy is the game Jenga, in which players take turns removing blocks from a tower and try to keep the increasingly narrow and unstable structure from toppling.)

Healthy ecosystems are complex and broad-based, made up of a number of species; this complexity makes for resilience—and a resilient ecosystem can better withstand threats such as acid rain or global warming. Make an existing ecosystem simpler and narrower, and you wind up with a more rigid and inflexible world, less able to withstand new threats.

If we continue at the current rate of deforestation and destruction of major ecosystems like rain forests and coral reefs wheremost biodiversity is concentrated—we will lose more than half of all the species of plants and animals on Earth by the end of the twenty-first century.

The rescue efforts in Vietnam are not happening in isolation but are part of a number of independent conservation projects that have sprung up, mushroomlike, over a broad swath of the tropics.

Bigger, scaled-up versions of tropical wildlife rescue projects like the EPRC may be the way to go. There is a precedent for this, in the form of the successful efforts of the National Tropical Botanical Garden to rescue endangered plants. Such work could be done at a very reasonable price; if the NTBG can do so much in Hawaii on $9 million per year, where land and expenses are high, imagine what could be done in Vietnam, where the average income is $1,100 per year. In short, conservation rescue work gets a lot of bang for the buck in tropical developing countries.

But at the moment, what will happen long-term in Indochina is hard to predict. Nadler and others say that countrywide, they are not very optimistic—but that when working on a species-by-species level, dealing with specific projects in specific locales, they see positive results, which gives them the hope to go on. And hope is important.

"What provides optimism is a local conservation hero—they inspire others to keep going with a given project," commented Alan Rabinowitz. "They may not think of themselves as heroes; they just have a need to do the work. But their efforts can continue even after they're gone."

He added, "Mind you, it's not about personal aggrandizement or ego. It's not about self. Instead, it's about being able to look around you and know that there are vast, beautiful areas out there that would not have been there otherwise, with more tigers, more jaguars, and more kinds of animals because of the work that's been done."

Asked why humanity ultimately should care if the natural, wild environment of a foreign place is destroyed, Rabinowitz said, "I live in New York, and I have yet to find a New Yorker who says we don't need Central Park. It is valuable real estate in the heart of a city that would be worth a fortune if it was ever developed—if it was converted to commercial real estate, the taxes would probably take care of our financial problems forever. But no New Yorker would dream of getting rid of it. It's a necessity for the city; it's the living, breathing heart. All of it could probably be re-created digitally somewhere, with fake scenery of trees and pumped in 'fresh' air—but people realize there's something more than that, something which we cannot artificially replicate."

So researchers return to Vietnam, year after year, to learn about the animals, their place in the environment, our place relative to them—and how to best protect all of it. They and their Vietnamese colleagues are attempting to put in place an ethic about the land that pioneering American environmentalist Aldo Leopold once explained as "A thing is right when it tends to preserve the integrity, stability, and beauty of the biotic community. It is wrong when it tends otherwise."

In making the effort to protect its natural heritage, we have learned more about the makeup of Vietnam—both good and bad—and the effort has shown us that Vietnam is a country, not a war. The rescue experience here has given good practice for the next tropical countries that open up: Cuba, Myanmar (Burma), North Korea, and the rest of the world's biological hot spots.

Meanwhile, the saola and muntjac rescue projects continue, as do efforts to control the black market. There is now a reserve for Cat Ba langurs, and a gibbon restoration project, in addition to the projects with turtles, cranes, butterflies, and numerous other species. A sea-turtle rescue project has started on the island of Con Dao in the south—the former Devil's Island of Indochina—so that a place that was once used to harbor prisoners is now home to a marine sanctuary. Eighty percent of the landmass of that island is protected, along with the marine sanctuary, and tourists can now spot endangered flowers sprouting amid the abandoned cells of the so-called "tiger cages" that once held human beings.

In the north, there's a project by Le Khac Quyet to rescue snub-nosed monkeys on the border with China. And at the EPRC, Nadler is steadily plugging away at restoring langur populations while the center continues to find new species on its doorstep—a "ferret-badger" appeared just a few months ago, as of the time of this writing.

Last I heard, Nadler had cut the speaker cord of yet another party that had disturbed the langurs' sleep. The Turtle Center continues to give classes on wildlife protection, and their teachers continue to down shot after shot of Vietnamese brandy with their graduating students on graduation night, as tradition calls for, giving their livers for science.

On a typical morning after, when the limestone mountains of Cuc Phuong emerge from the mist, lights come on in the windows of the farmhouses just outside the park; the buildings' traditional thatched roofs, combined with the adjacent neatly tilled rice paddies and abrupt nearby mountains, make the scene look as quiet and still as that on an ancient scroll. Inside the park, however, the forest is full of sound, from the drone of mosquitoes to the maniacal racket of white-crested laughingthrushes.

The bleary-eyed future park rangers, customs authorities, primatology students, rescue workers, volunteers, and others awake to the clamorous hoot of gibbons making their great calls, which manage to sound both sorrowful and beautiful at the same time.

Acknowledgments

There are some who can live without wild things,
and some who cannot.

—Aldo Leopold, author of *A Sand County Almanac*

Many people must be thanked for the creation of this book.

A big thank-you goes to my editors, Julia Pastore and Domenica Alioto of Crown Publishing, and to my agent, Matthew Carnicelli of Carnicelli Literary Management, for believing in those first rough outlines. I also owe a tremendous debt to David Bain of the Bread Loaf Writers Conference, who read those early magazine articles and recommended that I consider using the source material as the starting point for a book. I would also like to thank Nicholas Wade of the Science section of the *New York Times* for the time he took at his office at the paper one afternoon to offer suggestions on the manuscript-writing process.

Special thanks go out to taxonomist and mammalogist Colin Groves of the Australian National University (ANU) for pointing me toward what were to me at the time obscure field biologists in the middle of nowhere, and for his repeated help afterward. A big thank-you goes to Chris Bryant, founder of the ANU's Center for the Public Awareness of Science, and to the people at the Australian-American Fulbright Commission—who got me from America to Australasia in the first place, on a Fulbright Postgraduate Travelling Fellowship. It was more productive than I ever imagined, and a planned one-year stay became four as a result.

Among the field researchers, I must first and foremost express my gratitude to Tilo Nadler and his wife and interpreter, Hien, of the Endangered Primate Rescue Center in Cuc Phuong National Park, Vietnam. Their extensive hospitality, time, introductions to other researchers, help with logistics and officialdom, and warm support are deeply appreciated. I will long remember the delicious meals served at their home, surrounded by the rich souvenirs of a lifetime in the forests of Vietnam. I still have the crayon drawings given to me by their children, Khiem and Heinrich.

I am deeply indebted to the first US ambassador to Vietnam since the war's end, Douglas "Pete" Peterson, for making space in his busy embassy schedule to sit down with a freelance journalist traveling through Indochina. Special thanks goes to professor Vo Quy of the Center for Natural Resources and Environmental Studies at Vietnam National University for his background information on the founding of Vietnam's first national parks, and his tales of his own wildlife surveys conducted during wartime.

I also want to thank an entomologist, who shall remain unnamed, for demonstrating how much spectacular insect life can be collected at night. In addition, I want to say thanks to Isidore Rionduto for information on recording monkey vocalizations, and to Jeremy Phan for his guided tour of the back paths of the old-growth forest and the "Early Man Cave."

Also, thanks must go to Le Khac Quyet for sitting down with me across from Hoan Kiem Lake and giving information on his work with the Tonkin snub-nosed monkey; Howard Smith, editor of the *Vietnam News,* for tea, advice, and hints about coping with this non-Western culture; Nguyen Manh Cuong for his tour of plant and animal specimens at Cuc Phuong National Park; Bui Dang Phong and Tim McCormack of the Turtle Conservation Center; Tran Quang Phuong of the Small Carnivore Center (where I got to touch my first pangolin); Truong Quang Bich, director of Cuc Phuong National Park; and a big *"Cam on"* to Susan Rhind and Murray Ellis for their help on a few basic Vietnamese-language phrases and information on wild-animal husbandry.

I am grateful to *Los Angeles Times* foreign correspondent David Lamb, an "old Asia hand" par excellence, for his insights on Vietnam now and its contrasts with the war years when he was a reporter in Saigon, as Ho Chi Minh City was then called. Lamb is probably the only US newspaper correspondent from the Vietnam War era to later live in peacetime Hanoi, the former enemy capital.

I am grateful to Mike Hill of Fauna and Flora International for his information about the collecting of rare insects during our informal interview over a bowl of hot *pho* (noodle soup) in northern Vietnam. George Schaller, of Panthera, gave a wonderful sense of the big picture behind wildlife conservation—as well as what it was like for him when he discovered Vietnam's giant muntjac. Similarly, Alan Rabinowitz, also of Panthera, gave me a stronger idea of where conservation efforts in Vietnam fit in a global scale.

Nate Thayer's comments on his legendary expeditions in former Khmer Rouge territory gave me enormous insight into the difficulties of working in former battle zones. On a similar note, George Archibald of the International Crane Foundation helped me understand how, despite all appearances, wildlife can flourish in unexpected places, such as the demilitarized zone dividing the two Koreas. (At the time of our interview, Archibald had just returned from North Korea, departing that country less than an hour before the death of Kim Jong-il was announced.) Jeb Barzen and Tran

Triet, also of the ICF, gave me a better understanding of the nuts and bolts of environmental restoration in the Plain of Reeds.

Gert Polet and his wife, Ina, were vital to helping me understand the plight of the last rhinos in Vietnam while they were running the rhino program in Cat Tien National Park. I am also grateful to Ben Hodgdon of the World Wildlife Fund, and Dao Van Khuong, formerly of the Ministry of Agriculture Rural Development. Ben Rawson of Conservation International, and Nicholas Wilkinson of the Cambridge University Darwin Initiative Project, were invaluable for their perspectives on finding and protecting such creatures as the saola and the elusive (perhaps extinct?) kouprey. Vo Ngoc Thanh of the zoological department at Hanoi University was especially helpful in enlightening me about tigers in Vietnam.

Bert Covert of the University of Colorado at Boulder deserves special thanks for putting me in touch with young up-and-coming Vietnamese researchers likely to become leaders of the country's conservation efforts in the near future, and for steering me toward Van Long Nature Reserve.

Joe Walston of the Wildlife Conservation Fund proved an invaluable source for understanding what a working, functioning ecosystem is. Also helping me to get the big picture were Sue Lieberman of the World Wildlife Fund, and Sarah Halls of the International Union for the Conservation of Nature—both organizations headquartered in Gland, Switzerland, close to my then place of work in Geneva.

In New York City, Martha Hurley, biodiversity scientist at the American Museum of Natural History's Center for Biodiversity and Conservation and coauthor of 2006's *Vietnam: A Natural History,* gave me a priceless overview. Her colleague, Kevin Koy, of the center's remote sensing unit, was invaluable in explaining the process of mapping and census-taking.

My appreciation for taxonomy received a major boost after a behind-the-scenes tour of the Carl Linnaeus homestead in

Hammarby, Sweden, from Magnus Lidén, a curator of the botanic gardens at nearby Uppsala University. Likewise, my understanding of how tropical ecosystems preserve biodiversity was enriched with the aid of Jo Ann Valenti, Steve Perlman, Chipper Wichman, Gaugau Tavana, and the rest of the staff of the National Tropical Botanical Garden during a ten-day science journalism workshop at their facility in Kauai, Hawaii.

Mark Holdsworth and Steve Robertson of the Tasmanian Parks Service in Australia were invaluable in explaining the loss of the Tasmanian tiger and the psychology behind continued purported sightings; it helped me understand Vietnam's "cryptozoology."

Epidemiologist Jeanne Mager Stellman of Columbia University's Mailman School of Public Health gave a fascinating look into her National Academy of Science work, where her team conducted detective work that allowed for the creation of the most detailed computerized map ever made of herbicide spraying in Vietnam. Users of the map can now know how much Agent Orange and other defoliants were sprayed where and when and on whom.

Ngo Dzun of Thua Thiem Hue Tourism gave an excellent first-hand account of life in the DMZ between North and South Vietnam during the era of America's war in Vietnam.

Thanks to the Hanoi Backpackers Hostel and their staff for their help in getting me to the more remote places outside Sa Pa, their Internet access, and their unforgettable rooftop barbecues. It would be hard to come up with a friendlier transition to the north of Vietnam.

I am grateful to Boris Kolba of the National Writers Union for his input; Beth Livermore and her husband, Peter Hersh, for their advice and help; and all the people at the National Association of Science Writers, and at the American Society of Journalists and Authors. I also owe a big thank-you to my fellow ink-stained wretches at Bird by Bird—whom I must ask to forgive me for not being in touch more often. A special shout-out goes to Bill Burrows of my alma mater, New York University's Science, Health and Environmental Reporting Program.

I also appreciate the patience of my colleagues at CERN in Geneva, Switzerland, in letting me coordinate my time off to conduct interviews and research; thanks also to Vicky Huang, Simon Lin, and the staff of Academia Sinica in Taipei, Taiwan, for the tour of that city's famous Night Market.

Thanks also to the largely anonymous staffers who put together the excellent Charles Darwin exhibition at the Natural History Museum in London; the biggest ever of its type on the topic, it was assembled to celebrate Darwin's ideas for what would have been his 200th birthday. It was a thrill to see the original pages of Darwin's diary in his own handwriting and look at stuffed mockingbirds with the identifying tag COLLECTED BY C. DARWIN, GALAPAGOS.

Likewise, I want to thank the staff at the Lady Bird Johnson Wildflower Center at the University of Texas at Austin, who gave a wonderful behind-the-scenes tour for a group of environmental writers.

To Ernst Mayr, Alexander Agassiz Emeritus Professor of Comparative Zoology at Harvard University and its dean of biology for many years, I owe my thanks for kind words of encouragement. Although we had only a relatively brief interview and I felt somewhat tongue-tied at asking questions of "the pope of biology," the then ninety-six-year-old Mayr did much to help clarify some concepts. More important, he made me feel that an undertaking such as this book was possible. Now deceased, Mayr had a well-deserved reputation for taking an interest in the work of younger generations.

To Rebecca Meadows, I owe my gratitude for loaning me a quiet workspace outside Lausanne, Switzerland, in which to get my writing done during a critical phase; I also owe thanks to her cats, Cooper and Oscar, for not chewing to pieces the enticing piles of notes. Likewise, I must thank the long-suffering staff of the

Starbucks on the Rue Guillaume-Tell in Geneva, where so many hours were spent typing at their tables.

Over and above all else, I want to thank my parents, Barbara and Daniel Drollette Sr., for their patient support over the years of this project, especially when it felt at times as if it would never end.

Doubtless there are others I've forgotten, for which I must ask forgiveness. There are also those whose background information was invaluable but not put into print; even though their actual words may not have made it into the final text, their explanations and commentaries helped make this book deeper and more insightful than it could possibly have been otherwise.

As Charles Darwin wrote to his friend J. D. Hooker, "I thank you most sincerely for all your assistance; & whether or no my Book may be wretched you have done your best to make it less wretched."

Notes on Sources

AUTHOR'S NOTE

xi *"To see beneath both beauty and ugliness . . .":* T. S. Eliot, *The Use of Poetry and the Use of Criticism* (Cambridge: Harvard University Press, 1933).

PROLOGUE: DAWN IN THE JUNGLE

xv *"This region represents . . .":* Interviews with Colin Groves, July 1998.

xvi *"one of the wonders of the primate world":* Noel Rowe, *The Pictorial Guide to the Living Primates* (Charlestown, RI: Pogonias Press, 1998), 9.

xvi *"the unsung hero of Indochina wildlife protection":* Colin Groves interview, July 1998.

xvii *"a renaissance in species discovery . . .":* Virginia Morrell, "New Mammals Discovered by Biology's New Explorers," *Science* 273, September 13, 1996, 1491.

xvii *When mid–nineteenth century explorers . . .":* Paul Collins, "Gorillas, I Presume," *New Scientist,* October 1, 2005, http://www.cryptomundo .com/bigfoot-report/gorilla-timeline.

xviii *At times, it seems that everything is being sacrificed . . . "a miniature China on amphetamines":* Alan Rabinowitz, phone interview, January 11, 2012.

xviii *By 1812, noted French naturalist Georges Cuvier . . :* Martin J. S. Rudwick, *Georges Cuvier, Fossil Bones, and Geological Catastrophes: New Translations and Interpretations of the Primary Texts* (Chicago: University of Chicago Press, 1997), 207.

xix *"Our team's plane crashed . . .":* Nate Thayer e-mail interview, November 19, 2011.

xix *The jungle disease–resistant genes . . . could be worth billions:* Noel Vietmeyer e-mail, December 11, 2011. He repeated this comment many times previously, but it is probably most well known from Steve Hendrix's article "Quest for the Kouprey," *International Wildlife* 25, no. 5 (May 1995): 20–23.

xx *"When I'd ask local villagers . . .":* George Schaller, phone interview, December 19, 2011.

xxi **What Tilo Nadler did with just one creature . . . "unprecedented":** Colin Groves e-mail, October 24, 2000.

xxiii **Or "I don't want children, I need vigorous men":** Eric Thomas Jennings, *Imperial Heights: Dalat and the Making and Undoing of French Indochina* (Berkeley: University of California Press, 2011), 63–64.

CHAPTER 1: A PEACE MORE DANGEROUS THAN WAR

1 **"It's a Cruel, Crazy, Beautiful World.":** Johnny Clegg and Savuka, *Cruel Crazy Beautiful World,* Capitol, 1989.

1 **It had taken more than a year to determine:** Mark Kinver, "Javan Rhino 'Now Extinct in Vietnam,'" BBC News, October 25, 2011, http://www .bbc.co.uk/news/science-environment-15430787.

3 **"We identified a hundred and twenty different species . . .":** Field interview with Phil Benstead, November 13, 1998.

3 **"Parts of the park still have unexploded bombs and land mines":** Gert Polet interview. All first-person material from Polet comes from field interviews with him at his office in Cat Tien, November 13–15, 1998, and follow-up e-mails. In addition, the material on the park and its rhinos draws on his "Partnership in Protected Area Management: The Case of Cat Tien National Park, Southern Vietnam" (Hanoi: World Wide Fund for Nature, 1998). There is a shorter, more accessible PDF on the park called "Biodiversity Resources in Cat Tien National Park" (Hanoi: Ministry of Agriculture and Rural Development, 2012).

4 **The population of Indian rhinos:** Stephanie Pain, "Jungle Survivor: Vietnam's Unique Rhinos Are Facing Extinction," *New Scientist,* June 20, 1998, 25.

5 **"definitely a blow":** Mark Kinver, "Javan Rhino 'Now Extinct in Vietnam,'" BBC News, October 25, 2011.

5 **Alan Rabinowitz, a zoologist at the Bronx Zoo:** Bryan Walsh, "The Indiana Jones of Wildlife Protection," *Time,* January 10, 2008.

5 **"the dumbest thing":** Alan Rabinowitz, phone interview, January 11, 2012.

6 **Now, a little more than two decades later:** "Global Surge in Rhino Poaching," BBC News, December 1, 2009, http://www.news.bbc.co.uk/2/hi/science/nature/8388606.stm.

6 **"Consequently, rhino poaching is like the drug trade":** George Schaller interview, December 19, 2011.

CHAPTER 2: AN AMERICAN IN VIETNAM

11 **"'Tay ba lo' means 'Westerner with a backpack'":** Claude Potvin and Nicholas Stedman, *Do's and Don'ts in Vietnam* (Bangkok: Book Promotion and Service, 2008), 36.

12 *"We don't often like to mix that much with people":* Alan Rabinowitz, phone interview, January 11, 2012.

12 *"I knew of a couple . . .":* New York Times Magazine, June 19, 1994, 39.

12 *"Because orchid hunters hated the thought":* Susan Orlean, *The Orchid Thief* (New York: Ballantine, 2002), 59.

13 *Worse, if Marsh came across a dinosaur fossil:* "Dinosaur Wars," *American Experience,* http://www.pbs.org/wgbh/americanexperience/films/dinosaur.

13 *One of today's best-known paleontologists, Bob Bakker:* Ibid.

13 *"a two-legged monster armed with a gun . . .":* John James Audubon, *The Birds of America: From Drawings Made in the United States and Their Territories,* vol. 6 (New York: Roe Lockwood and Son, 1861), 19.

14 *But rather than throw out the internal organs:* Interview with Ernst Mayr at the 2002 annual meeting of the American Association for the Advancment of Science, Boston, Massachusetts. This weeklong meeting—the largest of its kind where thousands of scientists meet—is held in a different location each year in the United States. It is where major presentations are often made, such as the announcement of the deciphering of the human genome.

15 *"I've read about Tilo's primates . . .":* Alan Rabinowitz interview, January 11, 2012.

15 *"We never understood them . . .":* Stanley Karnow, *Vietnam: A History* (New York: Penguin Books, 1984), 19.

16 *"hard country":* Beth Livermore interview, New York City, February 1999.

16 *"Nate is simply the best reporter . . .":* Dale Keiger, "In Search of Brother Number One," http://www.jhu.edu/jhumag/1197web/brother.html.

16 *"After a series of furtive rendezvous . . .":* Nate Thayer, "Finding Pol Pot," http://www.publicintegrity.org/assets/pdf/pi_1999_03.pdf.

17 *"perhaps the most primitive of living cattle" . . . those figures were a good part:* Noel D. Vietmeyer, *Little-Known Asian Animals with a Promising Economic Future* (Washington, DC: National Academies Press, 1983), 53; e-mail interview and letter.

17 *"It's almost like the thing has some sort of ancient spell . . .":* Steve Hendrix, "Quest for the Kouprey," *International Wildlife* 25, no. 5, May 1995, 20–23.

18 *"These are very bad areas . . .":* Nate Thayer, "The Most Eccentric Hunt," http://birdlifeindochina.org/birdlife/report_pdfs/babbler_17.pdf.

18 *"It was Khmer Rouge territory . . . because nobody bothers them":* Nate Thayer interview, November 19, 2011.

19 *"I do not like this thing":* Conversation with Mr. Huyen, November 2, 1998.

20 For an extra fee, they could take some target practice: http://www
.easysaigontour.com/Cu_Chi_Tunnels.asp.

21 "Civil wars are always like that . . .": Jeb Barzen and Tran Triet interview, January 6, 2012.

22 a popular stereotype was: Robert Templer, *Shadows and Wind: A View of Modern Vietnam* (London: Abacus Books, 1998), 115.

22 wives of Northern generals filling up military planes: Karnow, *Vietnam: A History,* 37.

23 "Even though Nguyen isn't a formally trained scientist . . .": Jeb Barzen interview, January 6, 2012.

24 Vietnam now sees the United States: "Vietnam Embraces an Old Enemy," *International Herald Tribune,* August 29, 2011, 2.

24 "You fools! . . .": Karnow, *Vietnam: A History,* 153.

CHAPTER 3: A ROOM AT THE "HANOI HILTON"

26 "I still sometimes find myself thinking . . .": David Lamb interview, Hanoi, 1998.

26 the name Hoa Lo can be translated as "fiery furnace": Brochure handout at Hoa Lo Prison. For more information, see "Hoa Lo Prison," January 12, 2009, http://www.vietnam-beauty.com/cities/ha-noi/4-ha-noi/257-hoa-lo-prison.html.

27 I made a mental note: See the Red Cross website http://www.icrc.org/eng/war-and-law/treaties-customary-law/geneva-conventions/overview-geneva-conventions.htm.

28 "This head-cutting machine . . .": Handouts from visit to prison made on October 20, 1998.

28 "Guillotine Again Draws Gawking Crowds in Paris": Doreen Carvajal, "Guillotine Again Draws Gawking Crowds in Paris," *International Herald Tribune,* Weekend Arts, May 8–9, 2010, 16.

30 What to think of a culture: Snake Village advertising poster, March 18, 2012. An image of the poster will be available on the *Gold Rush in the Jungle* website, and can also be seen in the gallery for the *Gold Rush* Facebook page, http://www.facebook.com/Goldrush.in.the.Jungle/photos. References to the practice can also be seen in popular guidebook descriptions, such as those by Lonely Planet. Also, see http://www.gadling.com/2011/10/10/snake-village-in-hanoi-vietnam-allows-visitors-to-kill-and-eat.

31 "hours and hours of boredom . . .": Carolyn Webb, "POW's Journey to Australia, Via Love in Vietnam," *The Age,* September 17, 2009, http://www.theage.com.au/national/pows-journey-to-australia-via-love-in-vietnam-20090916-frp6.html.

31 "I had seen them at their worst . . .": Pete Peterson interview. Unless otherwise noted, all Peterson interviews were conducted at the US embassy in
Hanoi, October 27, 1998. There is a wealth of colorful material on Peterson
and his experiences as a former POW returning to Vietnam as US ambassador. A few standouts are David Lamb's book *Vietnam, Now* and Sandy
Northrup's PBS documentary *Pete Peterson: Assignment—Hanoi.*

32 "If I were an American voter" . . . "We would talk about the future . . .":
"Hanoi Hilton Jailer Says He'd Vote for McCain," *USA Today,* June 27, 2008.

33 "A lot of us have experiences here" . . . "You win by walking away":
Interview with Pete Peterson at US embassy in Hanoi, October 27, 1998.

33 "I want to heal the wounds . . .": Sandy Northrup (dir.), *Pete Peterson:
Assignment—Hanoi,* PBS, aired April 16, 1999, http://www.pbs.org/hanoi/.

34 Many of these veterans: Karnow, *Vietnam: A History,* 26.

34 "It was like growing up in a tough neighborhood": Author's 1998 journal
notes.

34 Some private organizations: BBC News, "The Legacy of Agent Orange,"
April 29, 2005, http://news.bbc.co.uk/2/hi/asia-pacific/4494347.stm.

34 "But since the end of the Vietnam War . . .": BBC News, "Vietnam's
War Against Agent Orange," June 14, 2004, http://news.bbc.co.uk/2/hi/
health/3798581.stm.

35 "In addition to effects on individuals . . .": Eleanor Jane Sterling, Martha
Maud Hurley, and Le Duc Minh, *Vietnam: A Natural History* (New Haven
and London: Yale University Press, 2006), 41.

35 "The United States and Vietnam have achieved . . .": US embassy Vietnam website, http://vietnam.usembassy.gov/ambspeech123010.html.

36 an American scientist doing field research on the effects of Agent Orange:
Interview with US Ambassador Peterson in his office, October 27, 1998;
interview with Jeanne Mager Stellman, December 1999.

36 Later, over coffee with a young Party member: Interview in Hanoi with
Nguyen Vu Tung, husband of ANU interpreter Tu Lan, October 21, 1998.

36 "They know as much about that war . . .": Charlie Rose, "A Conversation with Pete Peterson About Vietnam," *Charlie Rose,* PBS, aired July 31,
2001, http://www.charlierose.com/view/interview/3011.

37 "You cannot write about Vietnam . . .": David Lamb, *Vietnam, Now: A
Reporter Returns* (New York: Public Affairs, 2002), 3.

CHAPTER 4: ROUGHING IT IN RURAL VIETNAM

38 "Finding a new genus of mammal is always a shock": John Robinson, as
quoted in "Holy Bovid! A New Kind of Mammal Is Found in Vietnam,"
People, November 22, 1993.

38 At that time, Cuc Phuong was not yet in any guidebook: Sourcebook of Existing and Proposed Protected Areas in Vietnam, 2nd ed. (Hanoi: BirdLife International in Indochina, Vietnam Programme; Forest Inventory and Planning Institute of the Ministry of Agriculture and Rural Development, 2004).

39 After mistakenly greeting: Lonely Planet Vietnamese Phrasebook (Victoria, Australia: Lonely Planet, 2006), 99, 121; lesson from Tu Lan at the ANU.

42 In a country with a landmass about three-quarters the size of California: Eleanor Jane Sterling, Martha Maud Hurley, and Le Duc Minh, Vietnam: A Natural History (New Haven and London: Yale University Press, 2006), 3.

44 "There are about eighteen thousand tigers . . .": Interview with George Schaller, December 19, 2012. All Schaller material comes from this interview, conducted by phone during one of the few times this globe-trotting wildlife biologist was at his house in Connecticut. By his own account, he spends more time abroad than at home; he had just returned from India at the time of the interview.

44 "I would certainly be less of a fan of Tilo . . .": Joe Walston e-mail interview, November 2000.

45 "Zoos often become jails for animals . . .": Alan Rabinowitz interview, January 11, 2012.

45 "as much for the entertainment of the humans . . .": Thomas French, Zoo Story: Life in the Garden of Captives (Hyperion. New York, 2010), 80.

46 "fund-raising rackets": Nate Thayer e-mail interview, November 19, 2011. Thayer was one of the few to actually say this on record, as opposed to others who made the comment but did not wish to be quoted. For more about questionable wildlife practices, see Bryan Christy's story for National Geographic, "Asia's Wildlife Trade," January 2010. Also, Barry Yeoman's excellent story on bear farms, "Moon Bears in Distress, All for the Love of Bile," OnEarth magazine, December 6, 2011.

49 "Biology was possible to study . . .": Tilo Nadler e-mail, March 7, 2001. This quote was repeated to me in an e-mail on November 28, 2000, and during in-person visits in 1998 and 2010.

52 The World Conservation Union publishes: The IUCN Red List of Threatened Species, www.redlist.org.

54 It was a familiar story in developing countries: George Schaller interview, December 19, 2011.

54 "At the rate that the Delacour's langur . . .": Conservation International press release, August 25, 2004.

CHAPTER 5: A GRAND TOUR OF THE EPRC

59 *"Hope is important . . .":* Thich Nhat Hanh, *Peace Is Every Step: The Path of Mindfulness in Everyday Life* (New York: Bantam, 1992); Hanh is a peace activist and a Vietnamese Buddhist monk, now living in France, who was exiled from Vietnam.

61 *The first fast-food stand:* The Economist, "Fast Food in China: Here Comes a Whopper," October 23, 2008, http://www.economist.com/node/12488790.

62 *a young, low-level contact:* Interview with Vu Tung at Mocha Café, October 22, 1998.

63 *Miraculously, Indochina had continued:* Frank Zeller, "Illegal Wildlife Trade Takes Heavy Toll in Vietnam," Agence France-Presse, August 16, 2006; also see Eleanor Jane Sterling, Martha Maud Hurley, and Le Duc Minh, *Vietnam: A Natural History* (New Haven and London: Yale University Press, 2006).

63 *Many are "endemic":* Sterling, Hurley, and Minh, *Vietnam: A Natural History,* 1.

63 *Even some of the best-known species:* Ibid., 115.

63 *Vietnam's economy has been growing . . . lucky to grow about 3 percent annually:* CIA World Fact Book, https://www.cia.gov/library/publications/the-world-factbook/fields/2003.html; Indexmundi,http://www.indexmundi.com/vietnam/gdp_real_growth_rate.html; "Calculating When China Will Falter," *International Herald Tribune,* May 24, 2011, 19.

63 *a young and rapidly growing population:* Office of Statistics, *Annual Statistics of Vietnam,* http://www.gso.gov.vn/default_en.aspx?tabid=599&ItemID=9788.

63 *the average family has 6.7 children:* Tilo Nadler interview, October 25 and 26, 1998.

63 *Economists and financial pundits say:* "Is Vietnam the Next China?" *The Economist Blog,* May 20, 2011, http://www.economist.com/blogs/freeexchange/2011/05/reflections_hanoi_iii.

65 *he had two little holes in his foot as a memento:* Conversation with Nadler, Friday, March 19, 2010.

65 *On Mondays, there's rubbish strewn everywhere:* Nadler on-site interview, first day at EPRC, March 18, 2010.

68 *"They [douc langurs] are considered . . .":* Dee Ann Traitel, "Building a Safe Haven for Douc Langurs," ZooNooz (San Diego: San Diego Zoological Society, October 1996), 19.

68 *"A ship's crew putting in at Danang . . .":* Sterling, Hurley, and Minh, *Vietnam: A Natural History,* 227.

69 *Some scientists suspect that their song:* "New Gibbon Species Discovered in Indochina," http://idw-online.de/pages/de/news386717.

69 *"The female produces the 'great call' . . .":* Sterling, Hurley, and Minh, *Vietnam: A Natural History,* 112.

71 *by saving the Vietnam hot spot:* "Human Population in the Biodiversity Hotspots," *Nature,* April 27, 2000, 990–91.

73 *The langur succumbs to vomiting:* Interview with Jeremy Phan, March 21, 2010.

74 *"If you step back far enough,* **nothing** *matters":* Bill McKibben e-mail, December 16, 2011.

77 *Jim seldom had a line on-camera:* Mutual of Omaha's Wild Kingdom, Animal Planet, http://animal.discovery.com/fansites/wildkingdom/magnificent-moments/about/about.html and http://www.youtube.com/watch?v=T8s_g2v9M1g.

78 *"I didn't know why he needed to catch the anaconda . . .":* William B. Karesh, *Appointment at the Ends of the World: Memoirs of a Wildlife Veterinarian* (New York: Grand Central, 2000), 5.

CHAPTER 6: VIETNAM'S "LOST WORLD"

82 *the plants flourished so well:* Eleanor Jane Sterling, Martha Maud Hurley, and Le Duc Minh, *Vietnam: A Natural History* (New Haven and London: Yale University Press, 2006), 83.

82 *a record number that makes this country the most cycad-rich in Asia:* American Museum of Natural History, *Science Bulletin—Biodiversity Science in Vietnam,* http://www.amnh.org/explore/science-bulletins/bio/features/surveying-vietnam/biodiversity-science-in-vietnam. For more on cycads, see the 2009 *Gymnosperm Conservation at Cuc Phuong National Park,* by Nguyen Manh Cuong; the full report can be downloaded at http://www.ruffordsmallgrants.org/rsg/projects/cuong_nguyen_manh.

83 *Ninety miles of Son Doong have been mapped so far:* Mark Jenkins, "Conquering Vietnam's Megacave," *National Geographic* (January 2011), 114, 116.

83 *In the early 1940s:* Ibid.

85 *"just one big rice paddy":* Sterling, Hurley, and Minh, *Vietnam: A Natural History,*.

87 *Vietnam's most mountainous areas:* Ibid., 5.

88 *Consequently, many surveys in Indochina:* Ibid., 128.

88 *"A second later I saw another . . .":* Wilfred H. Osgood, *Mammals of the Kelley-Roosevelts and Delacour Asiatic Expeditions* (Chicago: Field Museum of Natural History, 1932), 208.

88 *colonial government–run opium monopoly:* Alfred W. McCoy, Cathleen

B. Read, and Leonard P. Adams II, *The Politics of Heroin in Southeast Asia* (New York: Harper and Row, 1972), 52, http://druglibrary.eu/library/books/McCoy/mccoy.pdf.

88 the French colony's opium business: Stanley Karnow, *Vietnam: A History* (New York: Penguin Books, 1984), 116–17.

88 More than ten thousand vascular plants are known: Sterling, Hurley, and Minh, *Vietnam: A Natural History,* 75.

89 By comparison, in our era: IUCN Red List website, http://www.iucnredlist.org.

90 "The Lang-Bian's hunting grounds . . .": Eric Thomas Jennings, *Imperial Heights: Dalat and the Making and Undoing of French Indochina* (Berkeley: University of California Press, 2011), 84–85.

94 "sly, hypersensitive, vengeful and capricious . . .": Helen and Walter Unger, *Pagodas, Gods and Spirits of Vietnam* (London: Thames and Hudson, 1997), 20.

95 One cave near Da Nang: Ibid., 142.

96 When I saw a traditional communal house: Nguyen Van Huy, Le Duy Dai, and Nguyen Quy Thao, *The Great Family of Ethnic Groups in Vietnam* (Hanoi: Bandotranthanh, 2009), 130–32; notes from visit to Vietnamese Museum of Ethnology.

96 For years, academics assumed: Keith Weller Taylor, *The Birth of Vietnam* (Berkeley: University of California Press, 1991), 313.

96 they say that the Dong Son developed it: "Southeast Asian Arts," *Encyclopaedia Britannica,* http://www.britannica.com/EBchecked/topic/556535/Southeast-Asian-arts/29458/Indigenous-traditions?anchor=ref402226.

103 "peculiar, varied and stressful physical conditions": Sterling. Hurley, and Minh, *Vietnam: A Natural History,* 18.

104 On the way, we see large numbers of butterflies: *Sourcebook of Existing and Proposed Protected Areas in Vietnam,* 2nd ed. (Hanoi: BirdLife International Vietnam Programme; Forest Inventory and Planning Institute of the Ministry of Agriculture and Rural Development, 2004), 2; http://birdlifeindochina.org/birdlife/source_book/source_book/Red%20River%20Delta/SB%20Cuc%20Phuong.htm and 2010 journal notes.

104 "a minor sensation": "New Gibbon Species Discovered in Indochina," *IDW Online,* September 21, 2010, http://idw-online.de/pages/de/news386717.

CHAPTER 7: AT THE HANDS OF THE DEMON CALLED "SCIENCE"

106 Something happens to the man . . .": Michael Crichton, *Congo* (New York: Random House, 1995), 247.

107 *His works were the precursor:* Susan Milius, "Biological Moon Shot: Realizing the Dream of a Web Page for Every Living Thing," *Science News* 173, no. 5, February 2, 2008, 72.

107 *"God created, Linnaeus arranged":* Robert Huxley, *The Great Naturalists* (London: Thames and Hudson, 2007), 133.

107 *"Taxonomy [the science of classification] is often undervalued . . .":* Stephen Jay Gould, *Wonderful Life: The Burgess Shale and the Nature of History* (New York: Norton, 1989), 98.

108 *"I do not know how the world could persist gracefully . . .":* Carl Linnaeus, *A General View of the Writings of Linnaeus: To Which Is Annexed the Diary of Linnaeus, Written by Himself, and Now Translated into English, from the Swedish,* Richard Pulteney and William George Maton, eds. (London: Cambridge University Press, 2011).

108 *"Love comes even to the plants":* Linnaeus, *Linnaeus: The Compleat Naturalist,* Wilfrid Blunt, ed. (Princeton: Princeton University Press, 2002), 33.

108 *Linnaeus even went so far:* "Animalia: Chordata," *Curious Expeditions,* http://curiousexpeditions.org/?p=31.

109 *"In addition to the first explanation . . .":* Gunnar Broberg, *Carl Linnaeus,* trans. Roger Tanner (Stockholm: Swedish Institute, 2006), 18.

109 *"disgusting names":* Huxley, *The Great Naturalists,* 134.

113 *"Nature, like society . . .":* Broberg, *Carl Linnaeus,* 30.

113 *"The question of exactly what constitutes . . .":* Eleanor Jane Sterling, Martha Maud Hurley, and Le Duc Minh, *Vietnam: A Natural History* (New Haven and London: Yale University Press, 2006), xiii.

114 *For instance, under one such system:* "Linnaeus," *Strange Science,* http://www.strangescience.net/linn.htm.

115 *"I had seen similar insects in cabinets at home . . .":* Alfred Russel Wallace, *The Malay Archipelago* (New York: Macmillan, 1886), 429.

115 *"Naturalists' Wall of the Dead":* Richard Conniff, *New York Times,* "Opinionator Blog," http://opinionator.blogs.nytimes.com/2011/01/16/dying-for-discovery.

CHAPTER 8: A BIOLOGICAL GOLD RUSH IN THE JUNGLE

116 *"The vilest scramble for loot . . .":* Joseph Conrad, "Geography and Some Explorers," *National Geographic,* March 1924.

119 *It was a new genus:* Malcolm W. Browne, "Scientists Hope More New Species Will Be Discovered in Vietnam," *New York Times,* May 3, 1994.

120 *"so strikingly unique . . .":* George Schaller interview, repeated in his July 1998 *International Wildlife* article, "On the Trail of New Species."

120 *the animal stood a good three feet tall:* Eleanor Jane Sterling, Martha Maud

Hurley, and Le Duc Minh, *Vietnam: A Natural History* (New Haven and London: Yale University Press, 2006), 13.

120 Although local people certainly knew: Sterling, Hurley, and Minh, *Vietnam: A Natural History,* 27.

121 "A lot of people were disappointed . . .": Colin Groves e-mail, November 22, 2006.

121 "The general is famous as a nature lover": Doug Fine, "The General and Me," *Washington Post,* February 11, 1996.

122 "a kinder, gentler dam": Jonathan Kent, "A Kinder, Gentler Dam," *Newsweek,* August 20, 2007, http://www.newsweek.com/2007/08/11/a-kinder -gentler-dam.html.

122 "Vietnam has a reputation for protecting . . .": George Schaller, phone interview, December 19, 2011.

123 "only slightly above eligibility for food stamps . . .": Harley Shaiken, University of California, Berkeley, as quoted in Vlasic and Bunkley, "U.S. Autoworkers Pushing for Higher Entry-Level Wage," *International Herald Tribune,* August 31, 2011, 18.

123 a 4,000 percent increase: Andrew Higgins, "China Runs Rings Around US Tariffs," *The Guardian Weekly,* June 3–9, 2011, 17.

124 "criminal networks have now shifted . . .": "Borderlines: Vietnam's Booming Furniture Industry and Timber Smuggling in the Mekong Region," Environmental Investigation Agency, http://www.eia-international.org/ borderlines.

125 The largesse gave him some political clout . . . "made me feel like I was covering a philanthropic entrepreneur in Cali or Medellín": Doug Fine, "The General and Me," *Washington Post,* February 11, 1996, http://www .dougfine.com/print/laos/.

125 The estimated global value of this trade: Duncan Graham-Rowe, "Endangered and In Demand," *Nature* (December 22–29, 2011), 101–3.

126 only about 3 percent intercepted: Frank Zeller, "Illegal Wildlife Trade Takes Heavy Toll in Vietnam," Agence France-Presse, August 16, 2006.

126 "empty forest syndrome": "Tigers on the Brink," World Wildlife Fund, http://www.wwf.dk/dk/Service/Bibliotek/WWF+i+Asien/Rapporter +mv./Tigers+On+The+Brink.pdf.

129 Many Asians believe: Quynh Anh, "Bears Tortured to Meet Asian Thirst for Bile," *Vietnam News,* October 3, 2009.

130 Its bile can be seen dripping: Bryan Christy, "Asia's Wildlife Trade," *National Geographic* (January 2011), 87.

131 The frighteningly thin bears are kept: Michael Bristow, "China Bear Bile

Farms Stir Anger among Campaigners," BBC News Online, February 29, 2012, http://www.bbc.co.uk/news/world-asia-china-17188043.

131 *"as a nation of animal abusers . . .":* Barry Yeoman, "Moon Bears in Distress, All for the Love of Bile," *OnEarth,* December 6, 2011.

131 *"Wildlife populations are dwindling . . .":* Zeller, "Illegal Wildlife Trade Takes Heavy Toll in Vietnam."

131 *"The use of satellite phones . . .":* Richard Thomas, "Huge Pangolin Seizure in China," TRAFFIC press release, July 13, 2010.

132 *Experts put the value:* Nadler interview, March 18, 2010.

132 *"the chicken soup of Chinese medicine":* Christy, "Asia's Wildlife Trade," 98.

132 *"So many well-educated people tell me . . .":* Nadler interview, October 25 and 26, 1998, p. 5

132 *seven rhino-horn smugglers:* Marie-Beatrice Baudet, "Even Dead Rhinos Aren't Safe," *The Guardian Weekly,* April 27, 2012, 30. For more on the boom in rhino horn, see Sarah Lyall, "Stuffed, Mounted and Dehorned—Poachers Targeting Rhinos in Museums to Satisfy Growing Demand in Asia," *International Herald Tribune,* August 25, 2011.

133 *Vietnam seems to have become the major epicenter:* Peter Gwin, "Rhino Wars," *National Geographic* (March 2012), 106–26.

133 *"There is no evidence at all . . .":* Raj Amin, interview, WNET/Channel 13, August 22, 2010, http://www.rhinoconservation.org/2011/03/29/busting-the-rhino-horn-medicine-myth-with-science/.

133 *"The makers of aspirin . . .":* Colin Groves's email, December 13, 2011.

134 *"We've seen your animals . . .":* Tilo Nadler interview. Unless otherwise noted, all Nadler interviews were conducted at the Endangered Primate Rescue Center, Cuc Phuong National Park, week of March 18, 2010.

134 *"The fact that you can get tiger meat . . .":* Edwin Wiek, quoted in "Illegal Wildlife Trade Takes Heavy Toll in Vietnam," *Agence France-Presse,* August 14, 2006.

134 *All of this is not to single out the Vietnamese:* Bryan Christy, "Asia's Wildlife Trade," *National Geographic* (January 2010), 92–93.

CHAPTER 9: THE KOUPREY—A CAUTIONARY TALE

137 *"Pol Pot killed two million people . . .":* Nate Thayer interview, November 19, 2011.

138 *Sauvel was well suited:* "Le Kou Prey ou Boeuf Gris Cambodgien"; *Bulletin of the Zoological Society of France* (June 8, 1937), 305–7.

139 *It may even have wandered off . . . circus escapees roaming:* Larry Collins

and Dominique Lapierre, *Is Paris Burning?* (New York: Simon & Schuster, 1965).

139 ***The animal is so elusive:*** IUCN Red List website and http://www.esabii .org/database/endangered/cetartiodactyla/bos_sauveli.html.

140 ***The animal could be important:*** Noel Vietmeyer, *Little-Known Asian Animals with a Promising Economic Future* (Washington, DC: National Academy Press, 1983), 53–59.

141 ***A few minutes of Wharton's rare footage:*** "In Search of the Kouprey," *A&E Investigative Reports,* http://www.youtube.com/watch?v=F-O56 _1lIQM&feature=mfu_in_order&list=UL.

142 ***"The most painful . . .":*** Steve Hendrix, "Quest for the Kouprey," *International Wildlife* 25, no. 5, September–October 1995.

142 ***any resulting progeny would have been:*** G. J. Galbreath, J. C. Mordacq, and F. H. Weiler, "Genetically Solving a Zoological Mystery," *Journal of Zoology* 270, no. 4 (2006), 561–64.

CHAPTER 10: COBRAS UNDER THE KITCHEN SINK

146 ***A group of about a dozen people:*** Author's 2010 journal, Friday, March 19.

148 ***Moths, it turns out, have their passionate admirers:*** Martin Wainwright, "Britain's Moths Enjoy Their Moment in the Spotlight," *The Guardian,* June 25, 2011, http://www.guardian.co.uk/environment/2011/jun/26/ moths-britain-varieties-martin-wainwright.

148 ***an annual "National Moth Night":*** "About Moth Night," http:// nationalmothnight.info.

148 ***websites devoted to the latest moth sightings:*** See, for example, "Atropos," http://www.atropos.info.

148 ***Tilo and the others start rapidly rattling off:*** Richard Seaman, "Bugs of Vietnam," http://www.richard-seaman.com/Insects/Vietnam/Highlights/ index.html.

150 ***Some suspect that they have:*** Eleanor Jane Sterling, Martha Maud Hurley, and Le Duc Minh, *Vietnam: A Natural History* (New Haven and London: Yale University Press, 2006), 112.

160 ***"I don't want any more of this crap . . .":*** Stanley Karnow, *Vietnam: A History* (New York: Penguin Books, 1983), 652.

CHAPTER 11: SURVIVING "AMERICAN WAR": AGENT ORANGE

162 ***Epigraph: "There was a special Air Force outfit . . .":*** Michael Herr, *Dispatches* (New York: Avon Books, 1978), 161–63.

163 *"Returning U.S. veterans now often find . . ."*: Michael Kelley, *Where We Were in Vietnam* (Ashland, Oregon: Hellgate Press, 2002), 7.

165 *slash-and-burn agriculture?:* Unless otherwise noted, all material here is from a tour of the old DMZ with Ngo Dzun and a visit to a Peace Village during November 1998.

166 *"War does not end . . ."*: Vo Quy speech in front of House Foreign Affairs Committee, June 4, 2009, http://foreignaffairs.house.gov/111/quy060409 .pdf. Additional information about Vo Quy's role as the father of Vietnam's environmental movement, and his thoughts about Agent Orange, can be found at http://www.terradaily.com/reports/Vo_Quy_father_of_Vietnams _environmental_movement_999.html.

166 *"Once dioxins have entered the body . . ."*: "Dioxins and Their Effects on Human Health," World Health Organization, Fact Sheet No. 225, http:// www.who.int/mediacentre/factsheets/fs225/en/.

167 *"the biggest remaining war-era issue"*: "US Helps Vietnam to Eradicate Deadly Agent Orange," BBC News, June 17, 2011, http://www.bbc.co.uk/ news/world-asia-pacific-13808753.

168 *"Studies suggest that this chemical . . ."*: "Agent Orange: Research on Health Effects of Agent Orange Exposure," US Department of Veterans Affairs, http:// www.publichealth.va.gov/exposures/agentorange/health_effects.asp.

169 *Elmo R. Zumwalt III was exposed:* "Elmo R. Zumwalt 3d, 42, Is Dead; Father Ordered Agent Orange Use," *New York Times,* August 14, 1988.

170 *Dr. Jeanne Mager Stellman is an epidemiologist:* All Stellman quotations come from interviews conducted at Stellman's office at Columbia University in December 1999.

170 *In the course of doing research to create the map:* "Flight Records Reveal Full Extent of Agent Orange Contamination," *Nature* (April 17, 2003), 649; "The Extent and Patterns of Usage of Agent Orange and Other Herbicides in Vietnam," ibid., 681–687.

172 *"fighting men comprised ten percent, or less . . ."*: Gerald Linderman, *The World Within War: America's Combat Experience in World War II* (New York: Simon and Schuster, 1997), 1.

CHAPTER 12: DMZ: THE THIN GREEN LINE

176 *"Animals don't need visas"*: Vo Quy interview, October 1998.

177 *"He protected Vietnam's flora . . ."*: Hannah Beech, "Heroes of the Environment, 2008: Vo Quy," *Time,* September 24, 2008,http://www .time.com/time/specials/packages/article/0,28804,1841778_1841782 _1841794,00.html.

179 *"In Vietnam, I once ate a pig . . .":* George Schaller interview, July 1998; George Schaller, "On the Trail of New Species," *International Wildlife* 28, no. 4, July/August 1998, 36.

181 *Analyzing the explorers' old journals:* Paul Martin and Christine Szuter, "War Zones and Game Sinks in Lewis and Clark's West," *Conservation Biology* 13, no. 1 (February 1999), 35–36.

182 *Similarly, Joseph Dudley and his colleagues:* Joseph Dudley et al., "Effects of War and Civil Strife on Wildlife and Wildlife Habitats," *Conservation Biology* 16, no. 2 (April 2002), 319–29.

182 *"War can be good . . .":* Hamish Clarke, "The Nature of War," *Cosmos,* May 9, 2007, http://www.cosmosmagazine.com/features/online/1289/the-nature-war.

182 *Up to ten feet wide:* The English Hedgerow Trust, http://www.hedgerows.co.uk.

183 *"two college kids on a farm . . .":* Interviews with George Archibald, founder of International Crane Foundation, and Jeb Barzen and Tran Triet of ICF; also see http://www.savingcranes.org.

CHAPTER 13: DRAGONS FLYING IN CLOUDS

186 *Epigraph: "The current destruction of our forests . . .":* Ho Chi Minh, quoted in Vo Quy, "Conservation of Forest Resources and the Great Biodiversity of Vietnam," *Asian Journal of Environmental Management,* 1992, 76.

187 *Even "Van Long" is colorful:* Dao Nguyen, "Van Long Nature Reserve," http://www.cenesta.net/icca/images/media/grd/van_long_vietnam _report_icca_grassroots_discussions.pdf.

187 *Located about halfway between:* Van Long interview notes and March 21, 2010, journal entries.

187 *As an added bonus:* Tilo Nadler, "The Status of Delacour's Langur (*Trachypithecus delacouri*) and the Possibilities for Its Long Term Conservation" (Frankfurt: Frankfurt Zoological Society, 2008), 49.

188 *The International Union for the Conservation of Nature puts the Delacour's langur:* "Primates in Peril," http://www.primate-sg.org/PDF/Primates.in.Peril.2008-2010.pdf, 49–50.

188 *The animal, named after one Jean Delacour:* Wilfred H. Osgood, *Mammals of the Kelley-Roosevelts and Delacour Asiatic Expeditions* (Chicago: Field Museum of Natural History, 1932), 195–98.

189 *The leaders of the local commune: Sourcebook of Existing and Proposed Protected Areas in Vietnam,* 2nd ed. (Hanoi: BirdLife International in Indochina, Vietnam Programme; Forest Inventory and Planning Institute of the Ministry of Agriculture and Rural Development, 2004).

190 *"There's no point in keeping them alive . . .":* Nadler on-site interview, October 25, 1998.

191 *The next step was releasing them:* Nadler e-mail interview, November 8, 2000.

191 *Those first few days afterward:* Nadler e-mail interview, March 7, 2001.

192 *Hopefully, at some point the animals' population thrives:* Jeremy Phan interview, March 21, 2010.

192 *"It used to be that you would . . .":* Nadler on-site interview, first day at EPRC, March 18, 2010.

193 *But after reading Lee Whittlesey's* **Death in Yellowstone:** Lee Whittlesey, *Death in Yellowstone: Accidents and Foolhardiness in the First National Park* (New York: Roberts Rinehart, 1995), http://www.yellowstonepark .com/moretoknow/shownewsdetails.aspx?newsid=41.

194 *Such positive local attitudes:* Nadler on-site interview, first day at EPRC, March 18, 2010.

194 *"This is a poor country . . .":* Nguyen Manh Ha interview, March 16, 2010

195 *Then we spot the first Delacour's langur:* All Van Long material comes from the author's notes from a visit on March 21, 2010.

195 *It is easy to understand:* Eleanor Jane Sterling, Martha Maud Hurley, and Le Duc Minh, *Vietnam: A Natural History* (New Haven and London: Yale University Press, 2006), 177.

196 *One early explorer:* Wilfred H. Osgood, *Mammals of the Kelley-Roosevelts and Delacour Asiatic Expeditions* (Chicago: Field Museum of Natural History, 1932), 205.

197 *This preserve is also well known:* http://www.greatvietnamholiday.co.uk/ locations/Van-Long-Nature-Reserve/103.

197 *This sensation of being in another world:* For more information about this remote reserve, see http://www.vietnamspirittravel.com/guide/van_long _nature_reserve_ninh_binh.htm.

198 *Such practices are strictly forbidden:* US Department of State, "U.S. Relations with Vietnam," http://www.state.gov/r/pa/ei/bgn/4130.htm.

CHAPTER 14: OF TURTLES AND SECOND CHANCES

203 *"There's four left in the world and only one of them is female":* Caroline Davies, "It's Love or Bust for Yangtze Turtles," *The Observer,* May 31, 2008, http://www.guardian.co.uk/environment/2008/jun/01/ endangeredspecies.animalbehaviour.

204 *"Tonkin cane bamboo":* http://thomasandthomas.com/rods/individualist; see also journals 2010 and 1998, Turtle Conservation Center interviews.

206 *Like many myths, it is part morality tale:* Harry Summers and Stanley Karnow, *Historical Atlas of the Vietnam War* (New York: Houghton Mifflin, 1995), 28.

208 *According to at least one biological reference source:* Eleanor Jane Sterling, Martha Maud Hurley, and Le Duc Minh, *Vietnam: A Natural History* (New Haven and London: Yale University Press, 2006).

209 *A guard was posted near the site:* Margie Mason, "Hanoi's Famous Hoan Kiem Turtle Is Captured to Receive Veterinary Care," *Los Angeles Times,* April 5, 2011, http://latimesblogs.latimes.com/unleashed/2011/04/hanoi-turtle-captured.html.

210 *A strong swimmer, this type of turtle:* Bui Dan Phong interview, March 18, 2010.

210 *long thought to be the only one of its kind:* www.asianturtlenetwork.org; Tim McCormack interview, March 18, 2010.

211 *The goal is for a low-impact environmental technology:* "High Tech Lake Clean-Up," *TerraDaily,* November 5, 2007, http://www.terradaily.com/reports/High_tech_lake_clean-up_to_save_Vietnams_legendary_turtle_999.html; "Turtle Awareness Campaign Launched," *Vietnam News,* March 23, 2010, 5; "It's Love or Bust for Yangtze Turtles," *The Observer,* June 1, 2008.

214 *"A little information helps":* Bui Dan Phong interview, March 18, 2010.

216 *In a May 2011 press release:* "Ancient Turtle Wants to Return to Hoan Kiem Lake," *InfoVietnam,* May 18, 2011, http://en.www.info.vn/science-and-education/nature-and-animals/26212-ancient-turtle-wants-to-return-to-hoan-kiem-lake-.html.

CHAPTER 15: WHY THE RHINO WENT EXTINCT IN VIETNAM

217 *"Understanding what a Vietnamese really means . . .":* Claude Potvin and Nicholas Stedman, *Do's and Don'ts in Vietnam* (Bangkok: Book Promotion and Service, 2008), 179.

219 *You need a minimum:* Mike Gilpin, "What Is the Smallest Number of Human Beings?" *Discover,* January 2001, 16.

220 *Other factors were items that Gert Polet alluded to:* Gert Polet interviews, November 13–15, 1998.

221 *Vietnam's population has boomed:* Maks Banens, Jean-Pascal Bassino, and Eric Egretaud, "Estimating Population and Labour Force in Vietnam under French rule (1900-1954)" (Asian Historical Statistics Project, Paul Valéry University, 1998); http://www.ier.hit-u.ac.jp/COE/Japanese/discussionpapers/DP98.7/98_7.html.

CHAPTER 16: VIETNAM'S YETI: WHAT ELSE IS OUT THERE?

225 *One of Kien's MIA team:* Bao Ninh, *The Sorrow of War* (London: Secker & Warburg, 1993), 88–89.

226 *A French colonist recorded spotting* l'homme sauvage *in 1947:* Vern Weitzel, *Nguoi Rung, Vietnamese Forest People, Wildman: Mythical or Missing Ape* (Canberra, Australia: Australia Viet Nam Science-Technology, 1998), http://coombs.anu.edu.au/~vern/wildman.html.

227 *a Yeti-like creature: Ape Story Lingers,* Vern Weitzel, ed. (Canberra, Australia: Australia Viet Nam Science-Technology, 1998), http://coombs.anu.edu.au/~vern/wildman/armyrpt70.txt.

228 *"This author's unit found . . .":* Michael Kelley, *Where We Were in Vietnam* (Ashland, Oregon: Hellgate Press, 2002), F-47.

228 *More recently, in September 2011:* Richard Freeman, "On the Trail of the Orang Pedak, Sumatra's Mystery Ape," *The Guardian,* Science Desk, September 8, 2011.

228 *"the largest professional, scientific and full-time organisation . . .":* Centre for Fortean Zoology, "Who We Are and What We Do," http://www.cfz.org.uk/.

229 *It started a comprehensive, systematic five-year search:* Lauren Gold, "Search Yields No Ivory Billed Woodpecker, but a Wealth of Ecological Information," *Chronicle Online,* July 15, 2009, http://www.news.cornell.edu/stories/July09/ivorybillsearch.html.

CHAPTER 17: TWO FUTURES: ANGKOR WAT OR KAUAI?

232 *"Biodiversity is lost thread by thread":* Chipper Wichman interview, Kauai, Hawaii, May 2007.

233 *"a popular university case study . . .": Caribbean Islands* (Singapore: Lonely Planet), 266.

235 *There are twenty specimens of* Hibiscus delphus: Interviews with researchers on the staff of the NTBG. For more on Hawaii's rare hibiscus species, see David Lorence and Warren Wagner, "Another New, Nearly Extinct Species of *Hibiscadelphyus* (Malvaceae) from the Hawaiian Islands" (St. Louis: Missouri Botanical Garden Press/Novon, 1995), 183; http://biostor.org/reference/64272. (Note, at the time, there were only four known specimens of *Hibiscus delphus.*)

235 *the only known wild specimen of the star violet:* William Laidlaw, *The Kadua Haupuensis: A Species in Transition* (Kauai, Hawaii: National Tropical Botanical Garden), http://people.ucalgary.ca/~laidlaw/haupuensis/haupuensis_web.html.

236 *We see almost a hundred of these:* Steve Perlman interview, May 2007; for more information about this trail, see http://www.kauaiexplorer.com/ hiking_kauai/awaawapuhi_hike.php.

238 *Such things do have a dollar value:* John Moir, "Calculating Nature's Invisible Benefits to Humans," *International Herald Tribune,* August 10, 2011, 15.

239 *governments such as that of Costa Rica pay landowners:* US Fish and Wildlife Service, *Valuing Ecosystem Services: Capturing the True Value of Nature's Capital* (Washington, DC: US Department of Agriculture, 2007), 2, http://www.fs.fed.us/ecosystemservices/pdf/ecosystem-services.pdf.

239 *Known as the Payment for Environmental Services Program:* Moir, "Calculating Nature's Invisible Benefits to Humans," 15.

240 *A single wild plant:* Center for Plant Conservation, *Cyanea Pinnatifida* (Honolulu: Harold L. Lyon Arboretum, 2010), http://www .centerforplantconservation.org/collection/cpc_viewprofile.asp?CPCNum =1188.

240 *Steve Perlman and Ken Wood found the only living example of the kanaloa plant:* David Berney interview; see also http://www.state.hi.us/ dlnr/dofaw/cwcs/files/Flora%20fact%20sheets/Kan_kah%20plant%20 NTBG_W.pdf.

240 *When the island was no longer needed for live-fire training:* Hanau hou he 'ula 'o Kaho'olawe: Rebirth of a Sacred Island (Wailuku, Hawaii: Kaho'olawe Island Reserve Commission), http://kahoolawe.hawaii.gov/history.shtml.

240 *Before the handover, a $400 million cleanup campaign was conducted:* For more information on the island's history, see http://www.oahu.us/ kahoolawe.htm.

243 *Perlman has been very successful:* http://plants.usda.gov/java/profile ?symbol=SIPE5.

243 *ahupua'a, or "stewardship of the land":* For more on ancient Hawaiian concepts of land division, see http://www.hawaiihistory.org/index .cfm?fuseaction=ig.page&CategoryID=299.

244 *Nā Pali Coast 'Ohana, a grassroots non-profit:* More on the work of this volunteer organization can be found at its extensive website, http://napali .org/.

CHAPTER 18: THE BIG PICTURE

246 *"Destroying rain forest for economic gain . . .":* E. O. Wilson, quoted in R. Z. Sheppard, "Nature: Splendor in the Grass," *Time,* September 3, 1990, http://www.time.com/time/magazine/article/0,9171,971049,00.html.

246 *"Everything we want, need, and use is dependent on nature":* George Schaller interview, December 19, 2011.

248 *"What provides optimism is a local conservation hero . . ."*: Alan Rabinowitz interview, January 11, 2012.

248 *"A thing is right when it tends to preserve the integrity . . ."*: Aldo Leopold, *A Sand County Almanac* (New York: Oxford University Press, 1949), 224–25.

ACKNOWLEDGMENTS

251 *"There are some who can live without wild things, and some who cannot"*: Aldo Leopold, *A Sand County Almanac* (New York: Oxford University Press, 1949).

253 *Lamb is probably the only US newspaper correspondent:* Allison Martin, "Interview with David Lamb," *Adopt Vietnam,* http://www.adoptvietnam .org/vietnamese/interview-lamb.htm.

257 *As Charles Darwin wrote:* Letter from Charles Darwin to J. D. Hooker, August 5, 1856, *Darwin Correspondence Project,* http://www.darwinproject.ac .uk/entry-1938.

Glossary

Who's Who Among the Animals

Annam flying frog *(Rhacophorus annamensis):* Much like a flying squirrel, this species glides from tree to tree. A large frog, it has webbing between its "hands" to help it glide.

Asian black bear *(Ursus thibetanus):* One of a number of bear species found in Indochina; it is sometimes anesthetized and its gallbladder repeatedly "milked" for the valuable bile, in a painful operation at so-called bear farms. Increasingly rare residents of tropical rain forest in Southern Asia.

Asian elephant *(Elephas maximus):* Indochina was once rife with elephants; the old name for Laos was "Kingdom of a Million Elephants." In Vietnam, they were used domestically for all sorts of things, including hauling troops into battle. Only about eighty or fewer survive in the wild in Vietnam.

Civet *(family Viverridae):* A type of nocturnal wildcat. This animal must take the prize for most unusual commercial uses: after raw coffee beans pass through its digestive system, the result is gathered up and processed into the most expensive gourmet coffee in the world, at an average of $300 per pound. Similarly, the animal has an anal gland that imparts a chemical that is highly desired by industry; the gland has long been the basis for some of the world's most expensive perfumes. (Something to think about the next time one sees expensive perfume in a shop.)

Edible-nest swiftlet *(Aerodramus fuciphagus):* Also sometimes known as Germain's swiftlet. As the name suggests, this species provides the main ingredient for the famous "bird's nest soup." The birds have a special gland under their tongue that enables them to secrete a viscous gluelike material that helps them build and attach nests to the sides of rock walls. The entire nest—which looks like a yellowy, translucent cup—is made of this saliva, without benefit of any grasses, twigs, or other supporting materials. As many as twenty million of the nests are harvested worldwide, with Vietnam having some of the highest-quality and most sought-after nests, commanding as much as $2,000 per pound in the Hong Kong Market.

Fishing cat *(Prionailurus viverrinus):* This cat has no fear of water. It likes to sit on the banks of streams and feed on crabs, fish, and mollusks, along with the usual mice. This gray-colored animal sometimes sits on a rock or a ledge over the water, dipping its paw in to scoop up small fish; at other times it simply plunges in.

Giant Mekong catfish *(Pangasianodon gigas):* Holds the record for world's largest freshwater fish, at 646 pounds, or about the size of a grizzly bear. As recently as fifteen years ago, about seventy of these fish were caught each year on the Mekong River in Cambodia. Since 2008, there have been no reported sightings of it.

Giant muntjac *(Muntiacus vuquangensis):* A kind of barking deer, with long, sturdy antlers. The giant muntjac was first found in 1994 by George Schaller; a new, smaller one was discovered in 1997.

Gibbon *(genus Hylobates/Nomascus):* A fruit-eater that looks somewhat like a very small ape with extraordinarily long arms; it lacks a tail. Famous for its eerie, deep-throated "great call," which Jane Goodall once described as "one of the wonders of the primate world."

Javan rhino *(Rhinoceros sondaicus annamiticus):* Long thought extinct on the Asian mainland, it was rediscovered in the forests of Cat Tien, a former Viet Cong staging point outside Saigon. As many as seven existed there in the 1990s; the latest report says that they were wiped out, with the last one killed by poachers.

Kouprey *(Bos sauveli):* Forest ox, weighing up to a ton. Probably the most monetarily valuable animal in terms of its genetic potential to the cattle industry. The only known living specimen in captivity, a calf, was held in a Paris zoo just before World War II. It died from starvation during the war, when food supply was short. Nevertheless, it is the official national mammal of Cambodia.

Langur *(subfamily Colobinae):* A type of monkey that feeds upon the leaves, seeds, and unripe fruit of tree species growing in limestone-rich environments. Its specialized stomach contains bacteria able to break down only these items; a langur cannot be fed any ripe fruit or commercial monkey chow, which makes it a difficult animal to house in a zoo. Likes to live in the trees and caves that pockmark the famous limestone mountains of Indochina's interior and the islands. Found in Southern Asia, from tropical rain forests to snow-covered regions. Prominent long tail. There are many species in this family, including the **black langur,** a species previously known only from a single skin collected in China in 1924. A few of the species recently found at the Endangered Primate Rescue Center in Cuc Phuong National Park include the **Ha Tinh langur,** the **Laos langur,** the **red-shanked douc langur**, the black-and-white **Delacour's langur**, and the **Cat Ba golden-headed langur** (discovered in 2000, and whose numbers are estimated at about 120 individuals).

Linh duong: Possibly a Vietnamese mountain goat. A specimen of this animal was reputedly shot on an expedition in the 1920s for use as tiger bait. Its bones were shipped to a Kansas museum, where they languished, forgotten, for the next sixty years. The horns of a newly killed one were later found in a traditional medicine shop in Ho Chi Minh City; they turned out to be fake.

Nguoi Rung: Literally, "Forest Man," this is the local Vietnamese villagers' term for a fur-covered creature that walks upright on two legs. Possibly apocryphal, or it may really exist. There is a report of it from the late 1960s, in the writings of Bao Ninh, a North Vietnamese Communist infantryman who was fighting along the Ho Chi Minh Trail.

Pangolin (*Manis pentadactyla*): Anteater-like creature with a long tail and scales that somewhat resemble those of an armadillo. Though it looks like a reptile, it is actually a mammal. This species has a chickenlike flavor that has made it highly sought after; in addition, a myth about the medicinal value of their scales makes them a target of the illegal wildlife trade.

Pygmy loris (*Nycticebus coucang*): Small (eight inches long), slow-moving, nocturnal animal that hunts insects, small reptiles, and birds. A closely related species is the even more lethargic **slow loris**.

Raccoon dog (*Nyctereutes procyonoides*): A tree-climber, native to Eastern Asia, and the only member of the dog family to hibernate in winter.

Saola (*Pseudoryx nghetinhensis*): A chocolate-colored, two-hundred-pound goatlike creature that is actually closest genetically to an ox. Prefers deep, wet, old-growth forests. In the late 1990s, ecologists estimated about a thousand of these shy creatures, with their long pairs of distinctive black horns, were living in the Annamite hills of Central Vietnam and Laos. The creature quickly became an icon for Vietnam's fledgling environmental movement. Despite its high profile, scientists at the World Wildlife Fund reported that in less than ten years saola numbers had crashed from about one thousand animals to around two hundred.

Sarus crane (*Grus antigone*): One of the most beautiful of the endangered cranes, which have recently made a comeback in the marshes of southern Vietnam. The cranes have a distinctive and elaborate dance, including a highly coordinated series of leaps, bows, and brief flights. Some researchers thought these dances were just for mating displays; however, sometimes entire flocks dance for no known reason.

Softshell turtle: This group of turtles contains species that are some of the rarest in the world, including the *Rafetus swinhoei*, of which only four are known to be in existence today. A specimen of it can be found in Hoan Kiem Lake (Lake of the Returned Sword) in the center of downtown Hanoi, where it is closely connected to a medieval legend about the saving of Vietnam from invaders. Turtles are traditionally one of Vietnam's four sacred animals.

Striped rabbit (*Nesolagus timminsi*): First showed up in a meat market on the Vietnam/Laos border in the 1990s; later, a live one was found just inside Vietnam's Pu Mat Nature Reserve. The species is thought to have originated in Borneo, nearly 1,200 miles away.

Tonkin snub-nosed monkey (*Rhinopithecus avunculus*): Named after the nearby Tonkin Gulf (as in the "Tonkin Gulf Resolution" of 1964). One of the country's largest monkeys, weighs up to thirty pounds.

Selected Bibliography

BOOKS

Banning, Jan. *Vietnam: Doi Moi*. Amsterdam: Focus, 1993.

Blunt, Wilfrid. *Linnaeus: The Compleat Naturalist*. Princeton: Princeton University Press, 2002.

Broberg, Gunnar. *Carl Linnaeus*. Stockholm: Swedish Institute, 2006.

Browne, Malcolm W. *Muddy Boots and Red Socks: A Reporter's Life*. New York: Crown, 1993.

Brunner, Bernd. *Bears: A Brief History*. New Haven: Yale University Press, 2007.

Burdick, Alan. *Out of Eden: An Odyssey of Ecological Invasion*. New York: Farrar, Straus and Giroux, 2005.

Burke, Andrew, and Justine Vaisutis. *Laos*. Singapore: Lonely Planet, 2007.

Clifton, Jon, and Jim Wheeler. *Bird-Dropping Tortrix Moths of the British Isles: A Field Guide to the Bird-Dropping Mimics*. London: Clifton and Wheeler, 2011.

Collins, Larry, and Dominique Lapierre. *Is Paris Burning?* New York: Simon & Schuster, 1965.

Corlou, Didier. *In the Heart of Hanoi: A Gastronomic Journey*. Hanoi: Tomorrow Media Co., 2008.

Cuong, Nguyen Manh, and Truong Quang Bich. *Gymnosperm Conservation at Cuc Phuong National Park*. Hanoi: Transportation Publishing House, 2009.

————. *Selected Common Plants at Cuc Phuong National Park*. Hanoi: Transportation Publishing House, 2009.

Darwin, Charles. *On the Origin of Species by Means of Natural Selection*. Foreword by Ernst Mayr. Cambridge: Harvard University Press, 2001.

Dash, Mike. *Tulipomania: The Story of the World's Most Coveted Flower and the Extraordinary Passions It Aroused*. New York: Three Rivers Press, 1999.

Desmond, Adrian, and James Moore. *Darwin*. London: Penguin Books, 1992.

Fountaine, Margaret. *Love Among the Butterflies: Travels and Adventures of a Victorian Lady*. London: Penguin, 1982.

Francis, Charles M. *A Field Guide to the Mammals of South-East Asia*. London: New Holland, 2008.

————. *A Photographic Guide to Mammals of South-East Asia*. London: New Holland, 2007.

French, Thomas. *Zoo Story: Life in the Garden of Captives*. New York: Hyperion, 2010.

Greene, Graham. *The Quiet American*. London: Penguin Books, 1977.

Guillon, Emmanuel. *Cham Art: Treasures from the Da Nang Museum, Vietnam*. London: Thames and Hudson, 2001.

Hai, Pham Hoang. *Vietnamese Water Puppetry*. Hanoi: VNA Publishing House, 2010.

Herr, Michael. *Dispatches*. New York: Avon Books, 1978.

Hochschild, Adam. *King Leopold's Ghost: A Story of Greed, Terror, and Heroism in Colonial Africa*. New York: Mariner Books, 1998.

Huxley, Robert. *The Great Naturalists*. London: Thames and Hudson, 2007.

Jennings, Eric Thomas. *Imperial Heights: Dalat and the Making and Undoing of French Indochina*. Berkeley: University of California Press, 2011.

Jones, John R. *Guide to Vietnam*. London: Bradt Publications, 1998.

Karesh, William D. *Appointment at the Ends of the World: Memoirs of a Wildlife Veterinarian*. New York: Grand Central Publishing, 2000.

Karnow, Stanley. *Vietnam: A History*. New York: Penguin Books, 1984.

Kelley, Michael P. *Where We Were in Vietnam: A Comprehensive Guide to the Firebases, Military Installations, and Naval Vessels of the Vietnam War, 1945–1975*. Ashland, Oregon: Hellgate Press, 2002.

Kemp, Hans. *Bikes of Burden*. Hong Kong: Visionary World, 2003.

Lamb, David. *Vietnam, Now: A Reporter Returns*. New York: Public Affairs, 2002.

Leopold, Aldo. *A Sand County Almanac*. New York: Ballantine Books, 1986.

Mann, Charles. *1491: New Revelations of the Americas Before Columbus*. New York: Vintage Books, 2006.

Matthiessen, Peter. *The Snow Leopard*. New York: Penguin Nature Classics/Penguin Group (USA), 1996.

Mayr, Ernst. *What Evolution Is*. New York: Basic Books, 2001.

McKibben, Bill. *The End of Nature*. New York: Anchor, 1989.

Mittermeier, Russell. *Primates in Peril: The World's 25 Most Endangered Primates, 2008–2010*. Bogotá, Colombia: Conservation International, 2008.

Nadler, Tilo. *Protected Animals of Vietnam*. Hanoi: Haki Publishing, 2008.

Ninh, Bao. *The Sorrow of War*. London: Secker & Warburg, 1993.

Orlean, Susan. *The Orchid Thief: A True Story of Beauty and Obsession*. New York: Ballantine Books, 2002.

Osgood, Wilfred H. *Mammals of the Kelley-Roosevelts and Delacour Asiatic Expeditions*. Chicago: Field Museum of Natural History, 1932.

Pariser, Eli. *The Filter Bubble: What the Internet Is Hiding from You.* New York: Viking, 2011.

Potvin, Claude, and Nicholas Stedman. *Dos and Don'ts in Vietnam.* Bangkok: Book Promotion and Service, 2008.

Quammen, David. *The Song of the Dodo: Island Biogeography in an Age of Extinction.* New York: Scribner, 1997.

Rabinowitz, Alan. *Jaguar: One Man's Struggle to Establish the First Jaguar Preserve.* Washington, DC: Island Press, 2000.

Ray, Nick, and Daniel Robinson. *Cambodia.* Singapore: Lonely Planet, 2008.

Ray, Nick, and Yu-Mei Balasingamchow. *Vietnam.* Singapore: Lonely Planet, 2009.

Rowe, Noel. *The Pictorial Guide to the Living Primates.* Charlestown, Rhode Island: Pogonias Press, 1998.

Salisbury, Harrison. *Vietnam Reconsidered.* New York: Harper & Row, 1984.

Santoli, Al. *To Bear Any Burden: The Vietnam War and Its Aftermath, in the Words of Americans and Southeast Asians.* New York: Ballantine Books, 1985.

Schaller, George B. *Tibet's Hidden Wilderness: Wildlife and Nomads of the Chang Tang Reserve.* New York: Harry N. Abrams, 1997.

Sjogren, Anna. *Linnaeus.* Stockholm: Hallgren and Fallgren, 2007.

Sterling, Eleanor Jane, Martha Maud Hurley, and Le Duc Minh. *Vietnam: A Natural History.* New Haven and London: Yale University Press, 2006.

Strange, Morten. *A Photographic Guide to the Birds of Southeast Asia, Including the Philippines and Borneo.* London: A and C Black Publishers, 2002.

Summers, Harry G., and Stanley Karnow. *Historical Atlas of the Vietnam War.* New York: Houghton Mifflin, 1995.

Taylor, Keith Weller. *The Birth of Vietnam.* San Francisco: University of California Press, 1991.

Templer, Robert. *Shadows and Wind: A View of Modern Vietnam.* London: Abacus Books, 1998.

Unger, Helen, and Walter Unger. *Pagodas, Gods and Spirits of Vietnam.* London: Thames and Hudson, 1997.

Van Huy, Nguyen, Le Duy Dai, and Nguyen Quy Thao. *The Great Family of Ethnic Groups in Vietnam.* Hanoi: Bandotranthanh, 2009.

Vietmeyer, Noel. *Little-Known Asian Animals with a Promising Economic Future.* Washington, DC: National Academy Press, 1983.

Vietnamese Phrasebook. Victoria, Australia: Lonely Planet, 2006.

Vietnam Travel Atlas. Victoria, Australia: Lonely Planet, 1996.

Weiner, Jonathan. *The Beak of the Finch.* New York: Vintage Books, 1995.

Wheeler, Tony, and Maureen Wheeler. *The Lonely Planet Story.* Hong Kong: Periplus, 2005.

Whittlesey, Lee. *Death in Yellowstone: Accidents and Foolhardiness in the First National Park*. New York: Roberts Rinehart, 1995.

Wilson, E. O. *The Future of Life*. New York: Knopf, 2002.

Zumwalt, Elmo R., and Elmo Zumwalt III. *My Father, My Son*. New York: Macmillan, 1988.

MAGAZINES, NEWSPAPERS, JOURNALS, AND DOCUMENTARIES

Anh, Quynh. "Bears Tortured to Meet Asian Thirst for Bile." *Vietnam News*, October 3, 2009.

Baudet, Marie-Beatrice. "Even Dead Rhinos Aren't Safe." *The Guardian Weekly*, April 27, 2012, 30.

BBC News. "US Helps Vietnam to Eradicate Deadly Agent Orange." June 17, 2011, http://www.bbc.co.uk/news/world-asia-pacific-13808753.

Butler, Declan. "Flight Records Reveal Full Extent of Agent Orange Contamination." *Nature*, April 17, 2003, 649.

Carvajal, Doreen. "Guillotine Again Draws Gawking Crowds in Paris." *International Herald Tribune*, May 8–9, 2010.

Christy, Bryan. "Asia's Wildlife Trade." *National Geographic*, January 2010.

Cincotta, Richard. "Human Population in the Biodiversity Hotspots." *Nature*, April 27, 2000, http://www.nature.com/nature/journal/v404/n6781/abs/404990a0.html.

Clarke, Hamish. "The Nature of War." *Cosmos*, May 9, 2007, http://www.cosmosmagazine.com/features/online/1289/the-nature-of-war.

Collins, Paul. "Gorillas, I Presume." *New Scientist*, October 1, 2005, http://www.newscientist.com/article/mg18825192.200-histories-gorillas-i-presume.html.

Coolidge, Harold Jefferson Jr. "The Indo-Chinese Forest Ox, or Kouprey." *Memoirs of the Museum of Comparative Zoology at Harvard College* 54, no. 6 (1940), http://ia700406.us.archive.org/19/items/memoirsofmuseumo5406harv/memoirsofmuseumo5406harv_djvu.txt.

Darwin, Charles. Letter from Charles Darwin to J. D. Hooker, August 5, 1856, *Darwin Correspondence Project*, http://www.darwinproject.ac.uk/entry-1938.

Davis, Mark (director). "Dinosaur Wars." *American Experience*, season 23, episode 3, 2011, http://www.pbs.org/wgbh/americanexperience/films/dinosaur.

Dudley, Joseph P. et al. "Effects of War and Civil Strife on Wildlife and Wildlife Habitats." *Conservation Biology* 16, no. 2 (April 2002), 319–29, http://onlinelibrary.wiley.com/doi/10.1046/j.1523-1739.2002.00306.x/abstract.

The Economist. "Fast Food in China: Here Comes a Whopper." October 23, 2008, http://www.economist.com/node/12488790.

The Economist Blog. "Is Vietnam the Next China?" May 20, 2011, http://www
.economist.com/blogs/freeexchange/2011/05/reflections_hanoi_iii.

Environmental Investigation Agency. "Borderlines: Vietnam's Booming Furniture Industry and Timber Smuggling in the Mekong Region," http://
www.eia-international.org/borderlines.

Fine, Doug. "The General and Me." *Washington Post,* Sunday Outlook, February 11, 1996, p. 1.

Freeman, Richard. "On the Trail of the Orang Pedak, Sumatra's Mystery Ape."
The Guardian, Science Desk, September 8, 2011, http://www.guardian.co
.uk/science/blog/2011/sep/08/orang-pendek-sumatra-mystery-ape.

Galbreath, G. J., J. C. Mordacq, and F. H. Weiler, 2006. *Journal of Zoology.* London, 2006. "Genetically Solving a Zoological Mystery: Was the Kouprey (*Bos
sauveli*) a Feral Hybrid?" 270 (4): 561–64.

Gilpin, Mike. "What Is the Smallest Number of Human Beings?" *Discover,* January 2001, 16.

Graham-Rowe, Duncan. "Endangered and In Demand." *Nature,* December 22–
29, 2011, 101–3.

Gwin, Peter. "Rhino Wars." *National Geographic,* March 2012, 106–26.

Hendrix, Steve. "Quest for the Kouprey." *International Wildlife Magazine,*
May 1995, 20–23.

Higgins, Andrew. "China Runs Rings Around U.S. Tariffs." *The Guardian
Weekly,* June 3–9, 2011, 17.

Jenkins, Mark. "Conquering Vietnam's Megacave." *National Geographic,* January 2011, 104–25.

Kurtis, Bill (producer). "In Search of the Kouprey." *A&E Investigative Reports,*
1999, http://www.youtube.com/watch?v=F-O56_1lIQM&feature=mfu_in
_order&list=UL.

Lyall, Sarah. "Stuffed, Mounted and Dehorned." *International Herald Tribune,* August 25, 2011, 1.

Martin, Paul S., and Christine R. Szuter. "War Zones and Game Sinks in Lewis
and Clark's West." *Conservation Biology* 13, no. 1, 35–36, http://www.jstor
.org/pss/2641562.

McDermid, Charles, and Cheang Sokha. "Search for the Kouprey." *Phnom Penh
Post,* April 21–May 4, 2006. Reprinted in *The Babbler: Birdlife International
in Indochina*: May 2006, http://birdlifeindochina.org/birdlife/report_pdfs/
babbler_17.pdf.

Moir, John. "Calculating Nature's Invisible Benefits to Humans." *International
Herald Tribune,* August 10, 2011, 15.

Northrup, Sandy (director). *Pete Peterson: Assignment—Hanoi,* PBS, aired April 16,
1999, http://www.pbs.org/hanoi/.

Office of Statistics. "Annual Statistics of Vietnam," http://www.gso.gov.vn/default_en.aspx?tabid=599&ItemID=9788.

Pigg, Susan. "French Set Tone at Hanoi Landmark." *Toronto Star,* January 20, 2007, http://www.thestar.com/comment/columnists/article/172089.

Polet, Gert. "Partnership in Protected Area Management: The Case of Cat Tien National Park, Southern Vietnam." Hanoi: World Wide Fund for Nature, 1998.

Remerowski, Ted, and Marrin Cannell (directors). *Legendary Sin Cities: Paris, Berlin & Shanghai.* Canadian Broadcasting Corporation, 2005.

Reporters Without Borders. "Internet Enemies: Vietnam," http://en.rsf.org/internet-enemie-vietnam,39763.html.

Roasa, Dustin. "Vietnam Cracks Down on Online Critics." *The Guardian,* World News, January 10, 2011, http://www.guardian.co.uk/world/2011/jan/10/vietnam-cracks-down-online-critics.

Rose, Charlie. "A Conversation with Pete Peterson About Vietnam." *Charlie Rose,* PBS, aired July 31, 2001, http://www.charlierose.com/view/interview/3011.

Sanchez-Azofeifa. "Costa Rica's Payment for Environmental Services Program—Intention, Implementation and Impact." *Conservation Biology* 21, no. 5 (March 21, 2007), 1165–73.

Stellman, Jeanne Mager. "The Extent and Patterns of Usage of Agent Orange and Other Herbicides in Vietnam." *Nature,* April 2003, 681–87, http://www.stellman.com/jms/Stellman1537.pdf.

Thayer, Nate. "Finding Pol Pot." *Newsletter of the Center for Public Integrity* 7, no. 2 (March 1999), 3.

———. "Freelancers' Vital Role in International Reporting." *Nieman Reports/Nieman Foundation for Journalism at Harvard,* Winter 2001, http://www.nieman.harvard.edu/reportsitemprint.aspx?id=101508.

Thomas, Richard. "Huge Pangolin Seizure in China." TRAFFIC press release, July 13, 2010, http://www.traffic.org/home/2010/7/13/huge-pangolin-seizure-in-china.html.

Thompson, Christian. "Tigers on the Brink." World Wildlife Foundation: Vientiane, Laos, 2010.

USDA Forest Service. "Valuing Ecosystem Services," http://www.fs.fed.us/ecosystemservices/pdf/ecosystem-services.pdf.

US Department of Veterans Affairs. "Agent Orange: Research on Health Effects of Agent Orange Exposure," http://www.publichealth.va.gov/exposures/agentorange/health_effects.asp.

Vittachi, Nury. "A Brief History of FEER." *The Curious Diary,* October 1, 2009, http://mrjam.typepad.com/diary/2009/10/a-brief-history-of-feer.html.

Vlasic, Bill, and Nick Bunkley. "U.S. Autoworkers Pushing for Higher Entry-Level Wage." *International Herald Tribune,* August 31, 2011, 18.

Webb, Carolyn. "POW's Journey to Australia, Via Love in Vietnam." *The Age,* September 17, 2009, http://www.theage.com.au/national/pows-journey -to-australia-via-love-in-vietnam-20090916-frp6.html.

Wheatley, Alan. "Calculating When China Will Falter." *International Herald Tribune,* May 24, 2011, 19.

World Health Organization: "Dioxins and Their Effects on Human Health." Fact Sheet no. 225, http://www.who.int/mediacentre/factsheets/fs225/en/.

Zeller, Frank. "Illegal Wildlife Trade Takes Heavy Toll in Vietnam." Agence France-Presse, August 16, 2006.

WEBSITES

Allwetterzoo Münster (http://www.allwetterzoo.de/englisch/index.php). Open-air zoo in Germany, devoted to showing wildlife in habitat that is as close to its natural environment as possible.

American Museum of Natural History (http://www.amnh.org). Renowned for its exhibitions and scientific collections, which serve as a field guide to the entire planet.

Asian Turtle Program (http://www.asianturtleprogram.org/project_page/ projects.html). Sponsors a number of projects focused on research, conservation, and awareness of Asia's most critically endangered tortoises and freshwater turtles, one of which is the turtle project at Cuc Phuong National Park.

Australian National Centre for the Public Awareness of Science (http:// cpas.anu.edu.au). Devoted to bringing science to all of the continent, it sponsors a traveling road show of science, called the "Science Circus," in which some of the contents of the national science museum are packed in a big truck, along with a dozen eager young scientists, who do a traveling road show of science in the outback. Think of it as *The Adventures of Priscilla, Queen of the Desert* meets Carl Sagan.

Bill McKibben (http://www.billmckibben.com). Author, educator, environmentalist, and scholar-in-residence at Middlebury College, Vermont.

Biodiversity Resources in Cat Tien National Park (http://www.cifor.org/ conservation/publications/pdf_files/vietnam/Cat_Tien_National_Park_ presentation.pdf). Contains some useful material, including maps, illustrations, and photographs, from Vietnam's Ministry of Agriculture and Rural Development.

BirdNet, Ornithological Council (http://www.nmnh.si.edu/BIRDNET/ orncounc/index.html). Website with comprehensive data about North American ornithological resources.

Cat Tien National Park (http://www.namcattien.org). Located in southern Vietnam, about an hour or two outside of Ho Chi Minh City, and home to the Javan rhino before it became extinct sometime in late 2011.

Center for Biodiversity and Conservation (http://cbc.amnh.org). The American Museum of Natural History created this interdisciplinary center in response to concern among its scientists over rapid species loss and increasing habitat degradation around the world.

Center for Fortean Zoology (http://www.cfz.org.uk). Self-proclaimed as the largest full-time organization in the world dedicated to cryptozoology—the study of unknown animals.

CIA World Fact Book (https://www.cia.gov/library/publications/the-world -factbook/fields/2003.html). Despite the name in the title, it is open to all.

The Colin Groves Pages (http://arts.anu.edu.au/grovco). Home of the mammalian taxonomist.

Columbia University Mailman School of Public Health (http://www .mailman.columbia.edu). Home of studies on the long-term effects of Agent Orange.

Con Dao Island (http://www.visitcondao.com). Former prison island, now the home of a large marine park and nature reserve.

Conservation International (http://www.conservation.org/Pages/default .aspx). Working "to ensure a healthy and productive planet for us all."

Convention on International Trade in Endangered Species (CITES) (http://www.cites.org). The organization's goal is to ensure that international trade in specimens of wild animals and plants does not threaten their survival.

Cornell Lab of Ornithology (http://www.birds.cornell.edu). Since 1915, this unit of Cornell University has worked to advance the study, appreciation, and conservation of birds. It is also home to the **Macaulay Library** (http:// macaulaylibrary.org), the world's largest and oldest scientific archive of biodiversity audio and video recordings, whose 175,000 recordings cover 75 percent of the world's bird species, along with a substantial percentage of insects, fish, frogs, and mammals.

Cuc Phuong National Park (http://www.cucphuongtourism.com). Site of the first national park in Vietnam and home of the Endangered Primate Rescue Center.

Cuc Phuong Turtle Conservation Center (http://www.asianturtlenetwork .org/project%20profiles/vietnam/cuc_phuong.htm). Site of many projects to save endangered turtles such as the softshell turtle in Hoan Kiem Lake.

Dan Drollette (www.dandrollette.com). Author's website; it contains his nonbook journalism and blog.

Darwin Initiative (http://darwin.defra.gov.uk). Site of a UK biodiversity program that assists countries that are rich in biodiversity but poor in financial resources.

Doug Fine: Author, Journalist, Adventurer, Goat-Herder (http://www.dougfine.com/the-premise). The title says it all.

Education for Nature Vietnam (http://www.envietnam.org). Vietnam's first non-governmental organization focused on the conservation of nature and the environment.

Encyclopedia of Life (http://www.eol.org). Global access to knowledge about life on Earth.

Endangered Primate Rescue Center (http://www.primatecenter.org/center.htm). One of the first wildlife rescue programs in Vietnam, it focuses on primates. Its methods have been copied and applied to programs devoted to saving other creatures.

Environmental Investigation Agency (http://www.eia-international.org). Independent organization dedicated to protecting the natural world from environmental crime and abuse.

Fauna and Flora International (http://www.fauna-flora.org). Protecting plants and animals through sustainable growth.

Florida Museum of Natural History (http://www.flmnh.ufl.edu). Good information on tropical plants and butterflies.

Frankfurt Zoological Society (http://www.zgf.de/?id=14&language=en). German non-profit committed to conserving biological diversity by preserving the world's natural environments.

Frozen Zoo (http://www.sandiegozoo.org/conservation/science/at_the_zoo/the_frozen_zoo). Keeps frozen cell cultures in support of worldwide efforts in research and conservation.

Gold Rush in the Jungle (www.goldrush_in_the_jungle.com). Website for the book, it contains audio and video clips of the animals mentioned, along with a blog, a gallery, extra material that could not fit in the printed text, and tips on travel in Vietnam. See also www.facebook.com/Goldrush.in.the.Jungle.

Ha Long Bay World Heritage (http://whc.unesco.org/en/list/672). In the Gulf of Tonkin in northern Vietnam, this preserve includes some 1,600 islands and islets, forming a spectacular seascape of limestone pillars.

Hanoi Backpackers Hostel (http://www.hanoibackpackershostel.com). Part of International Youth Hostels, this laid-back facility is in the center of the Old Quarter of Hanoi.

International Crane Foundation (http://www.savingcranes.org). Works worldwide to conserve cranes and the wetland and grassland ecosystems on which they depend.

IUCN Red List of Threatened Species (www.redlist.org). The definitive list of species, designed to determine the relative risk of extinction.

Kew Royal Botanic Gardens (http://www.kew.org). With its collections of living and preserved plants, of plant products and botanical information, it forms an encyclopedia of knowledge about the plant kingdom.

Lady Bird Johnson Wildflower Center, Austin, Texas (http://www .wildflower.org). Founded to protect and preserve North America's native plants and natural landscapes.

Linnean Society of London (http://www.linnean-online.org). Custodian of Linnaeus's collections, which comprises specimens of plants (14,000), fish (168), shells (1,564), and insects (3,198) acquired from the widow of Carl Linnaeus in 1784. Also contains his library and 3,000 of his letters and manuscripts.

Linnaeus Hammarby (http://www.hammarby.uu.se/LHeng.html). Home of Linnaeus, now a museum.

Linné on line (www.linnaeus.uu.se/online/about.html). A complete resource on the man known in Sweden as Carl Von Linné (and in English as Linnaeus). Translated directly from Swedish by his home institution, Uppsala University.

London Natural History Museum (http://www.nhm.ac.uk). Home of the new Darwin Center, opened to the public in September 2009.

Lord Howe Island (http://www.lordhoweisland.info). Home of the southernmost tropical reef in the world and located in the Pacific about a two-hour flight from Sydney, this Manhattan-sized island and World Heritage Site had zero contact with people until the first human settlers in 1834. By law, only 400 tourists may be on the island at a time, in addition to its 350 permanent residents. A naturalist's paradise, 70 percent of the island is a nature preserve.

Martin's Moths (http://martinsmoths.blogspot.com). "A fascinating insight into the secret lives of moths and men."

Millennium Seed Bank (http://www.kew.org/science-conservation/save-seed -prosper/millennium-seed-bank). Affiliated with Kew Gardens, its focus is saving seeds from plants faced with extinction. Working with a network of partners from fifty countries, they have successfully banked 10 percent of the world's wild plant species. The goal is 25 percent by 2020.

Missouri Botanical Garden (http://www.mobot.org). Sponsor of much field research and education on botany.

Museum of Cham Sculpture (http://www.vietnamtourism.com/e_pages/ country/province.asp?mt=84511&uid=811). Features artwork from the now-extinct India-based Champa or Cham civilization, whose glory days were from roughly the fifth through fifteenth centuries. Abandoned Cham buildings, temples, and fragments of cities dot central and southern Vietnam.

Nā Pali Coast 'Ohana (http://www.napali.org). Volunteer organization on the island of Kauai, it cleans up beaches and helps to restore and protect archaeological sites and the local environment.

National Association of Science Writers (http://www.nasw.org). Founded in 1934 as a forum in which writers join forces to improve their craft and encourage conditions that promote good science writing.

National Audubon Society (http://www.audubon.org). Named after John James Audubon, the society seeks to conserve and restore natural ecosystems, focusing on birds, other wildlife, and their habitats for the benefit of humanity and the Earth's biological diversity.

National Tropical Botanical Garden, Kauai, Hawaii (http://www.ntbg .org). Enriching life through discovery, scientific research, conservation, and education by perpetuating the survival of plants, ecosystems, and cultural knowledge of tropical regions.

Native Plant Information Network (http://www.wildflower.org/explore). Enter the name of any North American plant, and this network will give you a brief dossier on it.

On the Origin of Species by Means of Natural Selection (http://www .talkorigins.org/faqs/origin.html). The complete original text of Darwin's magnum opus, as it appeared in 1859, in electronic form.

Parks and Wildlife Service, Tasmania, Australia (http://www.parks.tas.gov .au). Proving that Tasmania is more than the Tasmanian Devil.

Phong Nha-Ke Bang National Park (http://whc.unesco.org/en/list/951). Protects a classic limestone karst environment in Vietnam.

San Diego Zoo (http://www.sandiegozoo.org). In addition to the zoo itself, the organization maintains one of the world's largest wildlife conservation programs.

Saola Working Group (http://www.savethesaola.org). Started by wildlife biologists in Vietnam who seek to find and protect these extremely rare species. One goal is to not only capture a live specimen for a prolonged period (it has only been done for up to a day or so) but have a pair of the animals successfully breed in captivity.

Sa Pa Homestay (http://sapahomestay.com). The folks to contact if you want to stay overnight with traditional indigenous tribespeople near the Vietnam/China border.

Science Health and Environmental Reporting Program (SHERP), NYU (http://journalism.nyu.edu/graduate/courses-of-study/science -health-and-environmental-reporting). One of the two oldest such programs in the United States.

Society of Environmental Journalists (SEJ) (http://www.sej.org). The premier site for environmental journalism.

Sourcebook of Existing and Proposed Protected Areas in Vietnam, 2nd ed. (http://birdlifeindochina.org/publication/Sourcebook). Produced by Birdlife International in Indochina, this "Source Book" contains a series of downloadable PDFs and PowerPoint slides designed for easy use by researchers and members of NGOs. Contains maps, images, and factoids on Vietnam's protected areas, especially its more remote and overlooked ones.

Temple of Literature, Hanoi, Vietnam (http://goseasia.about.com/od/vietnamstopattractions/a/temple_of_literature_hanoi.htm). This thousand-year-old temple, oldest in Hanoi, is a monument to the written word.

Thang Long Water Puppet Theatre, Hanoi, Vietnam (http://www.thanglongwaterpuppet.org/?/en/Home). One of the best places to see this traditional art form in Vietnam.

TRAFFIC Wildlife Trade News (http://www.traffic.org). Aims to ensure that trade in wild plants and animals is not a threat to the conservation of nature.

US Embassy—Hanoi, Vietnam (http://vietnam.usembassy.gov). A good source of current news on the area, in English.

Vietnam Embassy Online Newsletter (http://www.vietnamembassy-usa.org/news/story.php?d=20060719140601). Useful for travelers to Vietnam.

Vietnamese Cuisine/Didier Corlou (http://verticale-hanoi.com/en/la-verticale/didier-corlou). A member of the Culinary Academy of France, this Escoffier-trained chef has written numerous cookbooks and short guides to introduce Westerners to Vietnamese cuisine.

Vietnam Fine Arts Museum (http://www.vietnamonline.com/attraction/vietnam-fine-arts-museum.html). Located across the street from the Temple of Literature, this museum contains artwork going as far back as prehistoric and medieval Vietnam.

Vietnam Museum of Ethnology (http://www.vme.org.vn/aboutus_history.asp). Contains a rich collection of materials from Vietnam's estimated fifty-four different indigenous tribes, along with full-scale, working replicas of village meetinghouses, longhouses, stilt houses, bridges, and temples on its grounds.

Vietnam News (http://vietnamnews.vnagency.com.vn). The foremost English-language newspaper in Vietnam.

Voyage of the Beagle (http://www.thebeaglevoyage.com). Darwin's *Beagle* diaries presented as a blog, supplemented by maps and images.

War Remnants Museum (http://www.saigonscene.com/Museums.htm). Offers Westerners a chance to see, up close and personal, the Vietnam War from the opposing point of view. (The Vietnamese call it the American War.)

Wild Kingdom (http://www.wildkingdom.com/). One of the first reality TV shows, this program accompanied zoologists into the field as they caught wild specimens.

Wildlife Conservation Society (http://www.wcs.org). Saving wildlife and wild places across the globe, the society's story began in the early 1900s when they successfully helped the American bison recover on the Western Plains.

World Conservation Union (IUCN) (http://www.iucn.org). Confusingly enough, the letters stand for both "International Union for Conservation of Nature" and "World Conservation Union," depending on the language of a given country. By any name, this Swiss-based organization is the world's oldest and largest global environmental network, with 11,000 volunteer scientists in more than 160 countries.

World Wildlife Fund (http://www.worldwildlife.org). Their mission: to build a future in which people live in harmony with nature.

Index

About the Author

That's why journalism beckons so seductively to people
like Gallico and me; no other pursuit offers a
practitioner such richness of experience.

—Malcolm W. Browne, *Muddy Boots and Red Socks:*
A Reporter's Life

Dan Drollette Jr. is a science writer and editor—but above all else,
an avid nature lover. He first went hiking at age two, when his
parents put him in a backpack and climbed New York State's highest mountain (5,344 feet). In college, he spent two summers as a
part-time volunteer ranger/naturalist at Yellowstone.

He has written for peer-reviewed, scholarly journals but prefers writing for wildlife documentaries, nature magazines, and the science section of newspapers. To do so, he practices the "George Plimpton approach to science writing"—going into the field with his subjects to learn firsthand about the mud, sweat, and tears of their research.

Dan began as a photojournalist and held internships at *National Geographic* and the PBS series *Nature,* as well as several newspapers.

After earning a master's from NYU's Science, Health and Environmental Reporting Program, he received a Fulbright to Australia, where he wrote about science, technology, and environmental protection down under for audiences back home in the States. He continued to live and work in "Oz" as a freelance foreign correspondent for three more years, penning stories on everything from the physics of boomerangs to the migration of whale sharks. He has since traveled on assignment to Vietnam, Hawaii, South Africa, Sweden, Italy, Taiwan . . . and a bear den located behind a Massachusetts sausage factory.

Most recently, for three years he edited CERN's online computing magazine, outside Geneva, Switzerland.

For the past four years, Dan has been residing in Ferney-Voltaire, France, where he freelances as a science writer and editor for publications such as BBC World Wide. Samples of his work can be seen at his website, www.dandrollette.com, which also features his blog.

Those wishing to know more—including video clips, pictures, recordings of gibbons, the latest from the wildlife front, tips on traveling as a Westerner in Vietnam—can go to the website or to the Facebook page for the book: www.facebook.com/Goldrush.in .the.Jungle.

This is his first book.